ALL ABOUT
Science and TECHNOLOGY

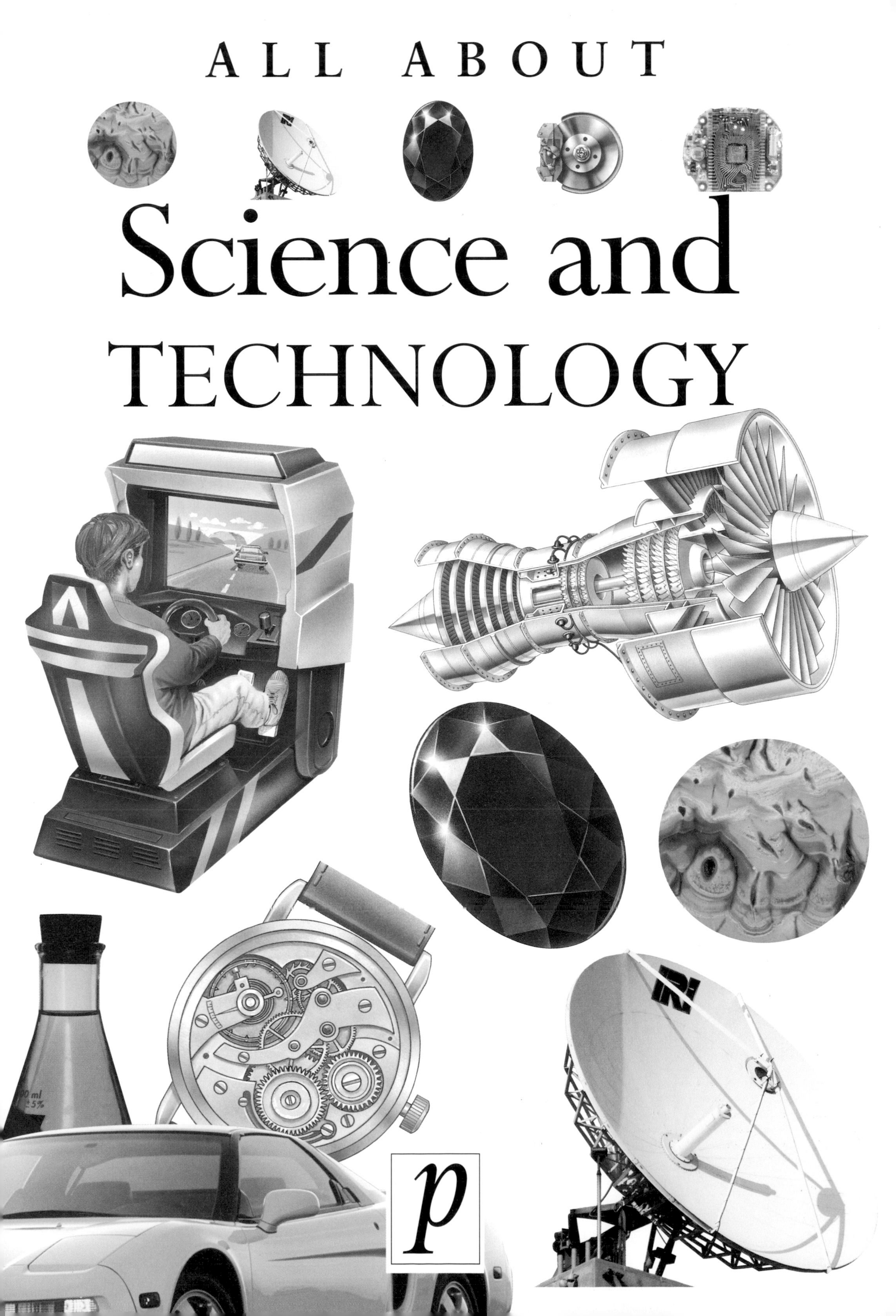
ALL ABOUT
Science and
TECHNOLOGY
p

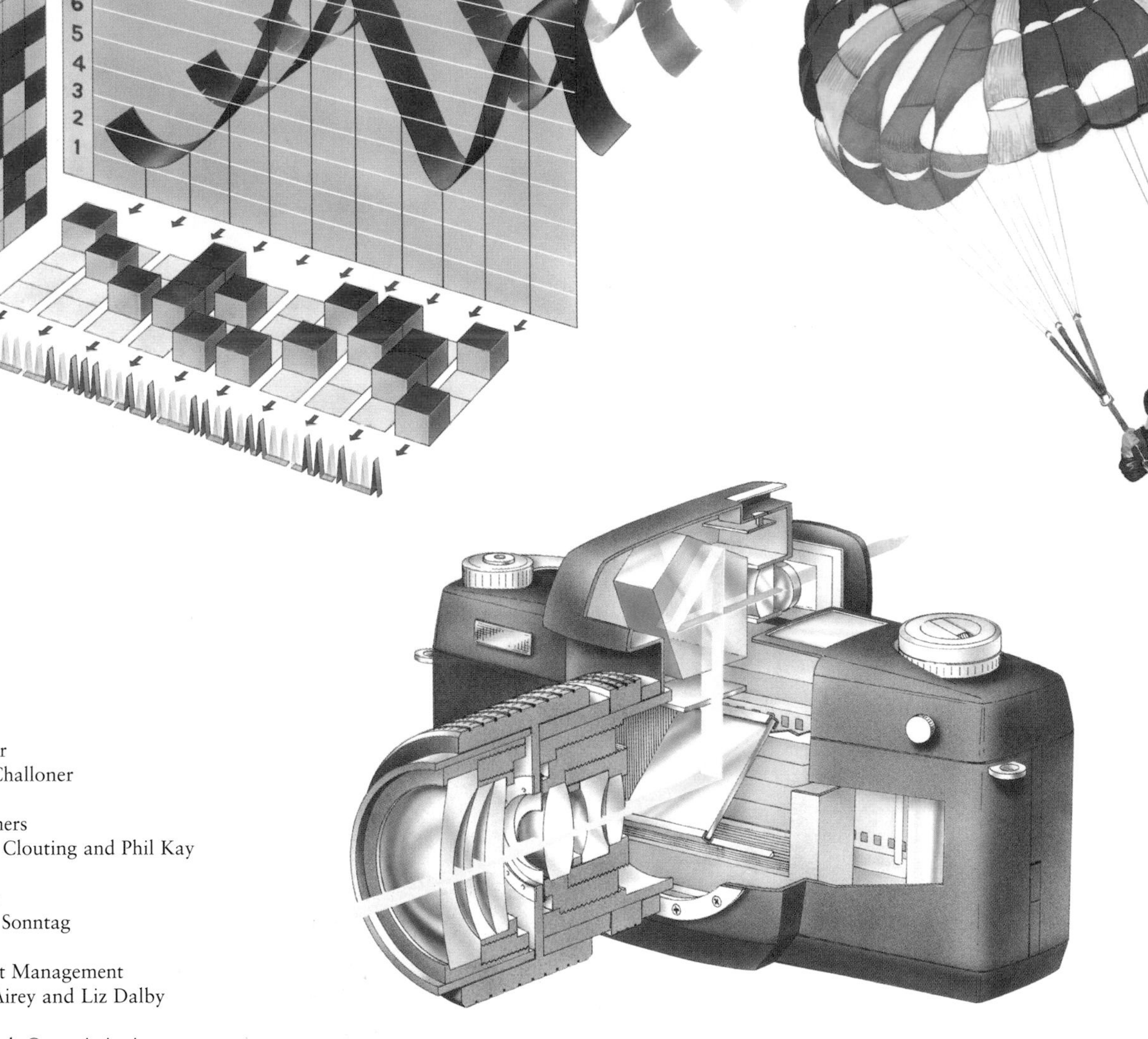

Author
Jack Challoner

Designers
Diane Clouting and Phil Kay

Editor
Linda Sonntag

Project Management
Raje Airey and Liz Dalby

Artwork Commissioning
Susanne Grant

Picture Research
Kate Miles and Janice Bracken

Additional editorial help from
Lesley Cartlidge and Libbe Mella

This is a Parragon Publishing Book
This edition published 2000

Parragon Publishing, Queen Street House, 4 Queen Street, Bath, BA1 1HE, UK

Produced by Miles Kelly Publishing Ltd
Bardfield Centre, Great Bardfield, Essex, England CM7 4SL

ISBN 0-75254-529-9

Printed in Singapore

Contents

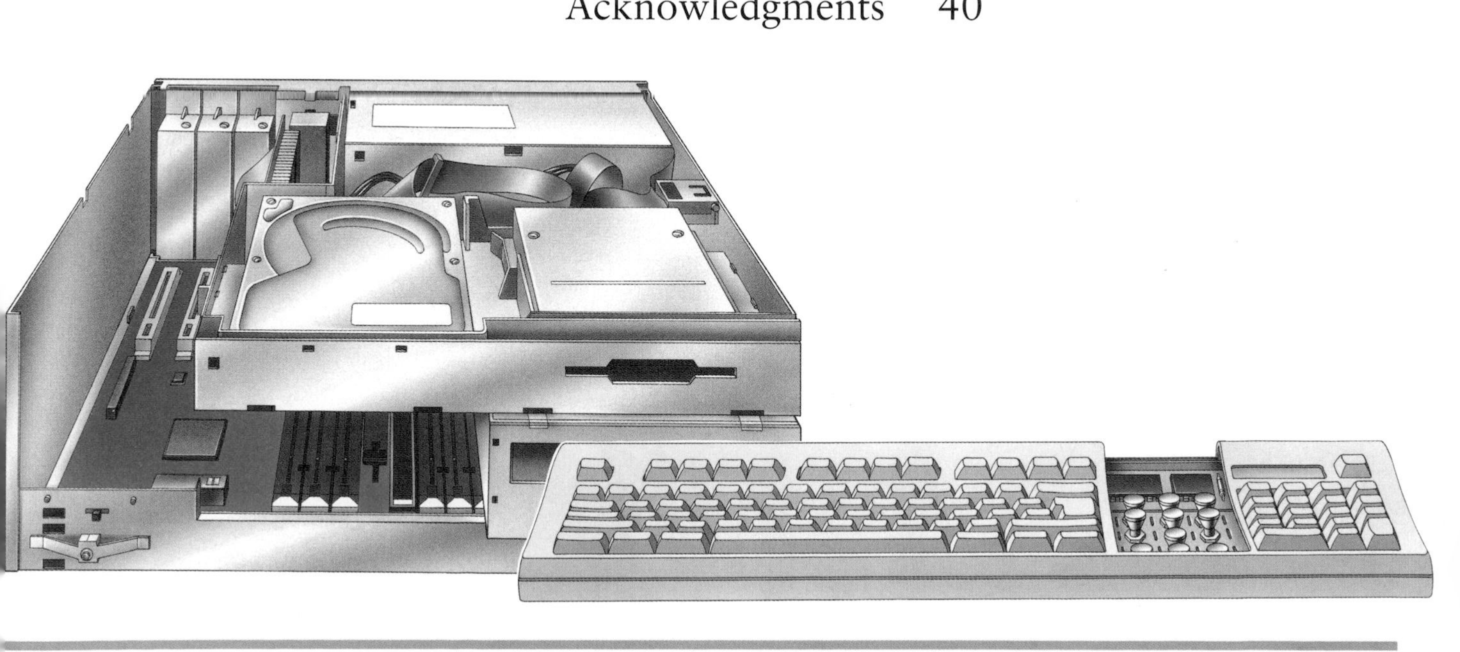

Science and Technology

All About Science and Technology is divided into fifteen different topics, each covered by a double page spread. On every spread, you can find some or all of the following:

- Main text to introduce the topic
- The main illustration, designed to inform about an important aspect of the topic
- Smaller illustrations with captions, to describe aspects of the topic in detail
- Photographs of unusual or specialized subjects
- Fact boxes and charts, containing interesting nuggets of information
- Biography boxes, about the scientists who have helped us to understand the way the world works
- Projects and activities

MATTER

STEEL, WATER, AND AIR are all examples of matter. Like most matter around you, they are made of tiny particles called atoms. Atoms are much smaller than the smallest speck of dust—a single one is far too small to see. About 90 different types of atom occur naturally. Substances made of only one type of atom are called elements. The gold of an earring, the copper of a water pipe, and the helium gas in a party balloon are all elements. Substances made of more than one type of atom are called compounds. In many compounds, atoms join together in groups called molecules. Water, for example, is a compound that consists of molecules. Each is made up of two atoms of the element hydrogen and one of oxygen, written as H_2O. A tiny drop of water consists of billions of these three-atom molecules. Most matter normally exists as one of three states: solid, liquid, or gas. Many solids are crystals, where the atoms or molecules are joined in a regular, repeating pattern.

Gases, liquids, and solids

A gas spreads out or expands to fill any space it can. The gases of air fill even the tiniest corners of a room. In a gas, the particles move randomly at high speed, bumping into each other and the walls of their container—like a small number of people blindfolded, running around a playground! If a gas cools down, the particles move more slowly, and come closer together. To be like a liquid, there would be more people in the playground, and they would walk around, occasionally linking arms. In a solid, the particles only move slightly around fixed positions. So the playground would be filled with people, linking arms and rocking to and fro on the spot! Most materials can exist in more than one state, depending on temperature. For example, liquid water freezes at 32 degrees Fahrenheit (0°C) into solid ice.

Everyday examples

These diagrams show the atoms or molecules in all three states of matter. The air that you breathe in is a mixture of several different gases, including the element oxygen and the compound carbon dioxide. Tea is a liquid solution that is a complex mixture consisting mainly of water. Steel is a solid that is a mixture of the elements iron and carbon, along with small amounts of other elements.

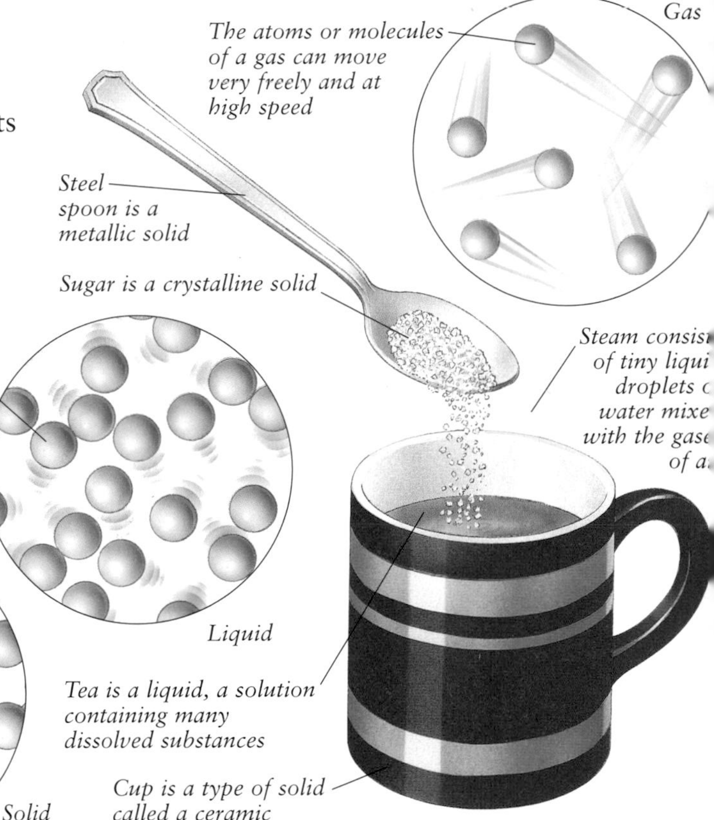

FACTS ABOUT MATTER

- The same elements that we find on Earth exist throughout the Universe.
- Plasma, a mixture of charged particles, is often called the fourth state of matter. On Earth, plasma is not common. Stars are made almost entirely of plasma, however.
- Pure hydrogen gas consists of molecules each made of two hydrogen atoms joined together, H_2. On average, a hydrogen molecule travels as fast as a speeding jet plane.
- Carbon has the highest melting point and boiling point of all elements. Helium has the lowest.
- Atoms are so small that it would take about 250 million of them in a line to measure 1 inch (25mm).

Elements, compounds, and mixtures

Pure substances may be elements or compounds. Pure water and pure salt are examples of compounds, whereas pure iron and pure hydrogen are elements. Most substances familiar from daily life are mixtures. Milk, for example, is a mixture of many compounds including water and fats. Adding sugar to water forms a type of mixture called a solution. The molecules of sugar break away from each other and mix with the water to become part of the liquid. Mixtures can be separated in many ways. A mixture of salt (a compound) and iron filings (an element) could be separated by adding it to water, to dissolve the salt, or by placing a magnet nearby to attract the iron. To separate mud from muddy water, you can pass the muddy water through a filter.

n atom of the element carbon

rotons (red) carry positive electric harge

Electrons (yellow) carry a negative electric charge

Neutrons black) carry o electric harge

here is the same umber of electrons round the nucleus as ere is of protons inside it

The nucleus has more than 99 percent of the mass of the atom

Hydrogen atom

Oxygen atom

Molecules of water

Different molecules

The simplest molecules consist of just two or three atoms. Substances called polymers consist of molecules made of hundreds or thousands of atoms. They are formed when smaller molecules join together. Plastics, such as polyethylene, consist of polymer molecules.

Molecule of polyethylene

Carbon atom

Hydrogen atom

/hat's inside?

toms are not the smallest particles of matter. Each atom onsists of even tinier particles called neutrons, protons, and ectrons. The neutrons and protons are together in the enter, or nucleus. Different elements have atoms with ifferent numbers of protons in their nuclei and different umbers of electrons going around the nucleus.

A regular arrangement

In a crystal, the regular way in which the atoms or molecules join together is the reason for the uniform appearance. Most crystals have a regular shape with flat sides or facets, angled edges and sharp corners. Crystals form as a liquid cools and solidifies, or as particles from a solution join together.

robing matter

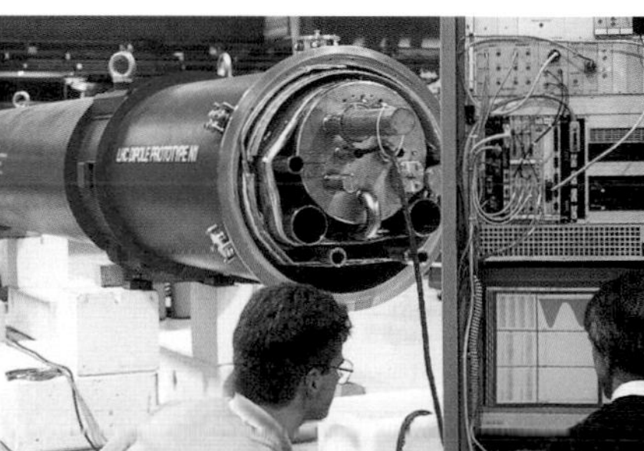

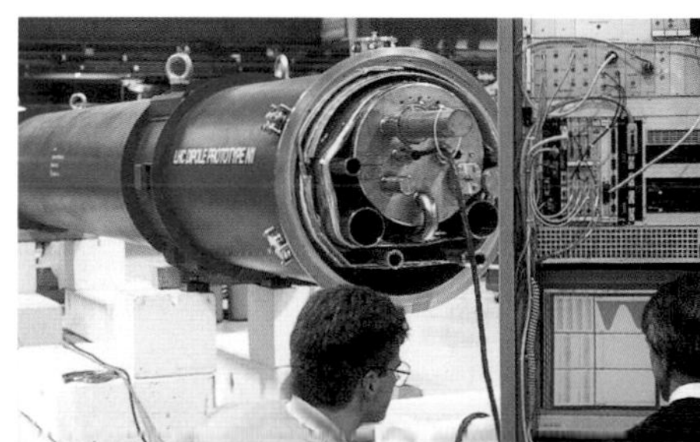

rotons and neutrons re not the smallest articles of matter. In evices called particle ccelerators, particles re made to collide at credibly high speed. howers of subatomic articles—those even naller than atoms—are produced in these ollisions. From the way these particles ehave, physicists are discovering more bout the matter that makes up not nly our whole world, but the whole the surrounding Universe.

ot liquid

/e normally think of the etal iron as a solid. At mperatures above ,795°F (1,535°C), iron elts to become liquid, or molten." It remains liquid ntil the temperature falls again, until it rises above 4,982°F ,750°C), when it becomes a gas!

Molten iron is liquid and pours easily, as in this steelworks

Growing Sugar Crystals

Ask an adult to heat a pint (450ml) of water in a saucepan, until it is hot but not boiling. Carefully add as much sugar as will dissolve in the hot water. Ask the adult to stir the solution, and then allow it to cool. Once the solution has cooled, pour it into the glass jar. Leave the jar somewhere where it will not be disturbed for a few weeks. The sugar molecules in the solution will gradually join up, forming beautiful crystals.

CHEMICAL REACTIONS

HYDROGEN GAS IS HIGHLY FLAMMABLE, and burns rapidly with a POP or even a loud explosion when mixed with air. Yet the result of this rapid, even violent burning is—water. Atoms of hydrogen (H) combine with atoms of oxygen (O) to make millions of tiny molecules of water (H_2O). The hydrogen and the oxygen are called the reactants, and the water is the product of the reaction. When a candle burns in air, the reactants are wax and oxygen from the air, while the products are water and carbon dioxide. Chemical reactions always make new substances from old ingredients, by rearranging the atoms and molecules of the reactants. For example, substances found in oil can be made to react in just the right way to make plastics such as polyethylene or nylon.

Three into two

Candle wax is made from just two elements: carbon (C) and hydrogen (H). When a third element—oxygen (O) from the air—gets involved inside the flame, atoms of the three elements rearrange, forming water (H_2O) and carbon dioxide (CO_2). Heat energy released by the reaction produces light and vaporizes more wax, ready to continue the reaction.

Wax is made up of huge molecules that contain the elements carbon and hydrogen

Oxygen gas is made up of molecules, each containing two atoms of the element oxygen

Water molecules are formed when some oxygen atoms join with hydrogen atoms

Carbon dioxide gas is formed when other oxygen atoms react with carbon atoms

Fast burner

Chemical reactions occur at different rates. An explosion is a very fast reaction. It produces gases so quickly that it causes a blast which can destroy buildings. This explosion was created by a reaction involving a material called dynamite.

Energy and reactions

All reactions involve a transfer of energy. In some cases, such as a burning reaction, energy is given out as heat or light. In others, the reaction needs energy to make it happen. It may take heat from its surroundings, for example. Such a reaction is found in a type of cold compress used to treat sporting injuries. Reactants inside the compress are mixed together when needed. They take heat they need to react from the injured part of the body. So the place where the compress touches the body is cooled. Another reaction that requires energy is used in photography. Silver compounds in photographic film produce pure silver to form the picture, but they need energy to do so. They get the energy from the light that falls on the film.

HENRY CAVENDISH (1731-1810)

French-born English physicist and chemist Henry Cavendish was the first person to identify hydrogen gas, which he called "inflammable air." Later, he worked out that water is produced when hydrogen and oxygen react together. He discovered this reaction at about the same time as it was discovered by the French chemist Antoine Lavoisier. The idea that water was made from hydrogen and oxygen contradicted the long-held belief that it was a pure element. Cavendish also devised a clever experiment to calculate the mass of the Earth. And he worked out how to calculate the forces between electric charges. Cavendish was an extremely shy man, had few friends, and worked mainly alone.

Electric reaction

All chemical reactions involve electricity, because they are caused by the exchange of the electrically charged particles called electrons, between atoms. On a larger scale, electric current can cause certain reactions to happen. This metal trophy has been coated by a layer of pure silver from the silver solution it was in, as an electric current passed through the solution. It is much less expensive than making the whole trophy from solid silver. The process is known as electroplating.

Billions of silver atoms have attached to the surface

Slow reaction

When oxygen and water mix with iron, a chemical reaction occurs. The iron slowly combines with the oxygen and water forming a reddish compound called hydrated iron oxide. This is more commonly known as rust.

Synthetic dyes

Chemists carry out research to invent or discover new chemical reactions that may be useful. About 150 years ago, the first synthetic or artificial dyes were produced by chemical reactions. This work has given us many more and brighter colored dyes than using natural dyes from plants and animal products.

Seeing the light

When light falls on a compound of silver on a photographic film, a chemical reaction occurs which frees, or liberates, silver metal from the compound. The silver forms tiny dark grains on the surface of the film that make up the photograph. The individual grains can only be seen under a microscope.

Black and blue

Many chemical reactions involve acids. Most metals, for example, fizz when placed in acids, dissolving to form a solution and giving off hydrogen gas. A solution of sulfuric acid dissolves the compound iron oxide, forming a blue solution of copper sulfate. If this solution is heated and then cooled, the copper sulfate forms into beautiful crystals.

FIZZY BUT SAFE

There are many simple and safe chemical reactions that you can perform at home or in school. One is the reaction between ethanoic acid, more commonly known as vinegar, and a compound called sodium hydrogen carbonate, more familiar as sodium bicarbonate (baking soda). A mixture of these two chemicals fizzes. Each of the bubbles of the fizz is filled with carbon dioxide gas produced by the reaction. To make the reaction look more impressive, add a few drops of green or red food coloring.

MATERIALS

AMONG THE MOST USEFUL MATERIALS in the modern world are wood, concrete, and steel for buildings, paper for writing, and printing, and glass for windows and bottles. Most people use the word "material" to describe substances like these, which we use to make things. To a scientist, however, a material is really any solid, liquid, or gas—even the air you breathe. Some materials come from plants or animals or from underground. Examples of these natural materials are paper, leather, and marble. Other materials are made by chemical reactions in factories. Metals are examples of such "synthetic" materials. They are obtained from rocks in which they are chemically combined with other elements. Plastics are also synthetic materials, because they are made when compounds in oil undergo carefully controlled chemical processes. Scientists are always striving to find new materials for an ever-increasing variety of products.

Properties of materials

The way in which a material behaves helps us decide how to use it. Glass, for example, is transparent and this makes it suitable for windows. Textiles are soft, opaque and easy to cut to size. This makes them perfect for keeping out light as curtains or blinds. The quality, or property, of a material can be changed by mixing it with other materials. When metals are mixed together, they are called alloys. Some alloys are stronger than individual metals and may have been produced for a specific purpose. Another way in which materials are combined is in composites. An example of a composite is fiberglass, which is a combination of plastic and glass, used to make small boats.

There's so much in it

Modern consumer products are often made from a bewildering array of different materials. The people who made this games console used glass, metals, plastics, and fabrics. The plastic materials have only beome available in the last hundred years, while glass and metals have been used for many centuries.

Special materials called phosphors, painted on the transparent glass screen, glow to make the picture

Silicon, copper, and gold are the main materials in the electronic circuits

Foam and fabric give the seat the desired comfort

Heat is used to mold rigid plastic into shape for the seat back

Sturdy steel base

KNOW YOUR MATERIALS

- The first plastic was celluloid, which was used to make photographic film.
- The first metal to be used was gold. This is because it occurs naturally, whereas most other metals must be smelted to remove them from rocks called ores.
- Whereas water molecules consist of just three atoms each, plastics have very large molecules, each one of which consists of thousands of atoms.
- Rubber is a natural material—it is solidified latex, which is the sap of a rubber tree. Its properties can be changed to make it stronger and more elastic, by a process called vulcanization. This involves heating rubber under pressure with sulfur. The process can also turn rubber into a harder substance, called vulcanite.

What's it made of?

It may surprise you to know that glass is made of sand. In a glassworks, sand is mixed with other substances, and heated until it melts. As it cools, it becomes thicker, and can be molded or blown (right) into shape before it hardens to form glass. Metal-containing compounds may be added to make colored glass.

The ingredients of glass, from the top, are calcium carbonate, sand, and sodium carbonate

Together they are strong

Concrete can hold up a load that would squash other materials. That is why it is used for the pillars of this suspension bridge. Steel is strong if you try to pull it. That is why it is used to make the cables, which are under incredibly high tension. This winning combination allows engineers to build bridges with long spans.

lloy there!

mixture of two metals, or a metal and another material, is called ı alloy. Steel, for example, is an alloy of iron, carbon, and a ıriety of other elements. The body, or fuselage, of jet airliners is ɔrmally made from alloys of aluminum. These alloys are ideal ɛcause they are light yet strong.

gain and again

'e use millions of tons of materials to make things every day. 'hen their useful life is over, some of these things are buried in landfill sites, where they slowly decay. Others can be recycled—the materials from which they are made are processed again and made into new things. Empty drinks cans are separated into those made of steel and those made of aluminum, before being melted down.

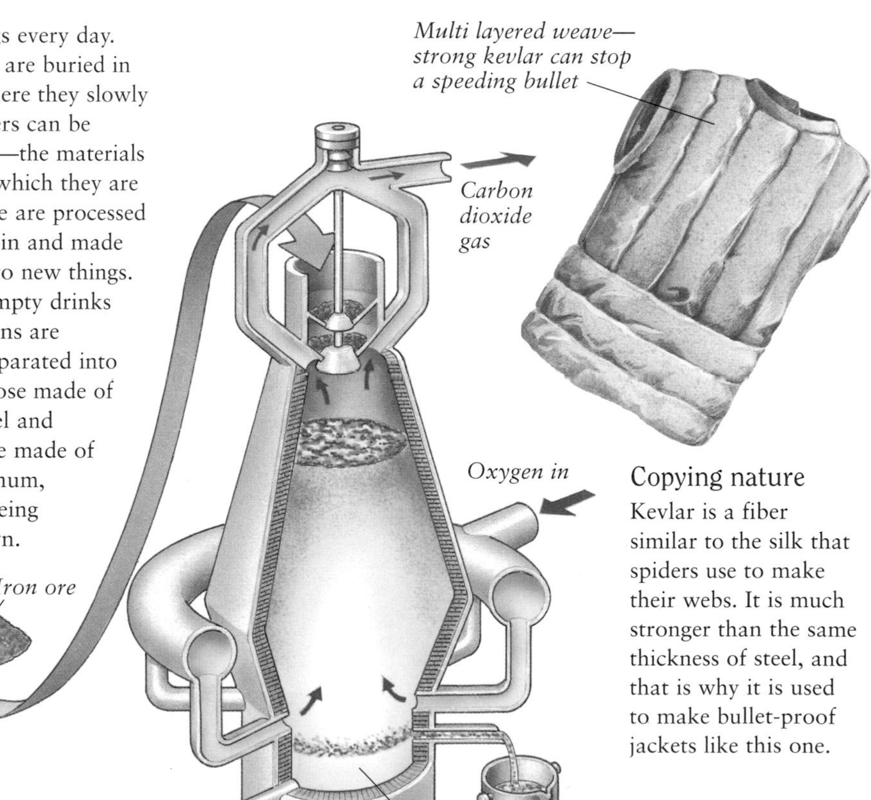

A blast furnace

Copying nature

Kevlar is a fiber similar to the silk that spiders use to make their webs. It is much stronger than the same thickness of steel, and that is why it is used to make bullet-proof jackets like this one.

's a steel

ɔn is made from iron ore, fairly common mineral und all over the world. 'hen iron is combined ith carbon and a small nount of other chemical ɛments, it makes steel, hich is more durable an pure iron.

ENERGY

SCIENTISTS DEFINE ENERGY as "the ability to do work", in other words, to make things happen. The Sun supplies most of the Earth's energy, as electromagnetic radiation that includes light rays and infrared or heat rays. Some of the light energy is absorbed by plants, and stored in their leaves and stems as energy-rich chemicals such as sugars. When we eat plants, some of the energy is released inside our bodies, to help us grow and stay alive.

Energy is never created. It merely changes from one form to another. For example, the moving, or kinetic, energy of the wind turns a wind turbo-generator, which converts it into electrical energy. This energy can make light in light bulbs, heat in a toaster, or movement in an electric motor.

Electricity can also be made from plants and animals that died millions of years ago. As they rotted, they formed fossil fuels—coal, oil, and natural gas. The energy of those plants and animals is still contained in the fossil fuel, which is burned in power stations. It forms heat that is used to generate electricity.

Power to the people

Most of the electricity that we use is made, or "generated," in power stations. The electrical energy is distributed to where it is needed by long, thick metal cables. These are usually high above the ground, for safety, and supported on towers called pylons. The network of cables crisscrossing the landscape is known as the electricity distribution grid.

Energy transfer

Energy from the Sun is absorbed by plants by a process called photosynthesis. When a cow eats grass, it takes in this chemical energy. The cow uses some to grow and stay alive, and stores some in its body. When a person eats the cow's meat or drinks its milk, he or she takes in some of the energy from the cow. And some of this energy is converted again when a person runs, into kinetic energy. To stop, the person uses friction between the running shoe and the ground, producing heat energy.

Plant captures light energy from the sun by the process of photosynthesis

Cow eats plant and takes in its energy as sugars and other substances

Person eats burger and takes in cow's meat with its stored energy

Energy in food is changed by muscles into energy of movement

Energy needs

Most of our energy needs are supplied by burning fossil fuels. These fuels can be burned and used only once. Scientists have shown that supplies of fossil fuels are being used up extremely fast. The race is on to harness safe and low-cost "alternative" energy sources, such as solar, wind or water power. These alternatives are "renewable." For example, the kinetic energy of the wind is produced by the Sun heating different parts of the Earth's surface and so causing the air to move. More of this energy is released every day by the Sun, so wind power will never run out. Another alternative is nuclear power, produced in nuclear power stations by a reaction called fission. This is the controlled splitting of the atomic nuclei of "heavy" elements such as uranium, and it releases huge amounts of heat energy. This can be used to generate electricity. However, there are hazards involved in fission, including the release of harmful radioactivity.

The larger the blades, the more energy captured

Wind power

Wind energy has been used for centuries, in windmills that use the energy to grind grain. Today, we use wind turbines to turn generators that supply electricity. They rely on strong winds, which only blow steadily in certain places. Wind power stations known as wind farms, with lots of wind turbines, are usually built in high, remote sites. Unlike fossil fuel power stations, they do not produce any air pollution. But they alter the look of the landscape, which has been called "visual pollution." They may also make loud humming or whooshing noises.

Hot or cold

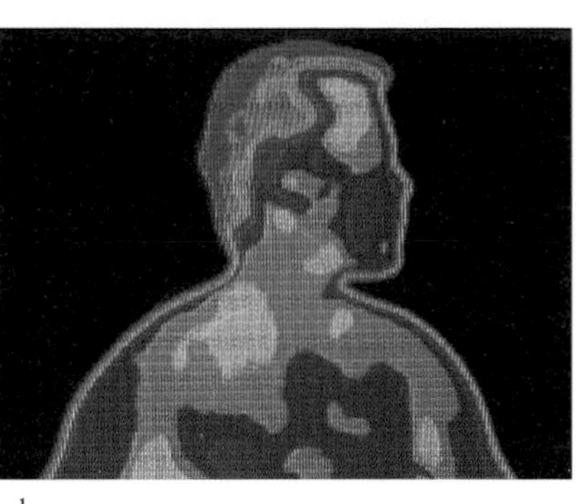

When things feel cold to the touch, it is because they are at a temperature which is lower than that of your skin. This is normally 98.6°F (37°C), the average temperature of the human body. But it varies slightly around the body, as shown by this thermograph or computer-colored heat-photograph. Warmer parts of the body are shown in redder colors, and cooler parts such as the scalp and ears in blues and greens. Temperature is a measure of the internal energy of an object's atoms and molecules – how much and how fast they move. In a hot solid object, the atoms and molecules vibrate more and faster than when the object is cool. Heat energy transfers from hot things to cold things in three different ways. These are called conduction, convection and radiation.

Future energy?

Inside this experimental fusion reactor, the nuclei from the centres of hydrogen atoms join, or "fuse," to form the nuclei of helium atoms. (This is the same process that happens inside the Sun.) Nuclear fusion releases huge amounts of energy, and produces virtually no waste, pollution or radioactivity. But at present, the process requires more energy input than it produces.

Up and down

When you climb up to a high ridge, you need energy, to work hard as you ascend against the pull of gravity. The energy is not lost: it can be retrieved by jumping off the ridge! The energy which comes from a certain position, being higher than you were, is known as potential energy. It becomes kinetic or moving energy, as you jump and speed downward. This energy is then stored in the bungee rope as it stretches, in the form of potential energy. It is retrieved again when the rope pulls and bounces you back up!

Potential energy is stored in the stretchy rope

Wind turbines swivel on their towers, always to face into the wind

Generator inside the turbine produces electricity as the blades spin

Wires in towers link turbines to wind farm control room

Control room

FORCES AND MOTION

A FORCE IS A PUSH OR A PULL. We can observe how forces can change motion by considering everyday objects such as shopping carts. A cart moving through outer space will not speed up, slow down, or change direction, because no force acts on it. Back on Earth, a cart outside a supermarket will not move unless someone pushes or pulls it. A full trolley is much harder to start and stop, and to steer, than an empty one. English scientist Isaac Newton observed these same effects, as detailed below. He described them as the Laws of Motion (even though he never used a shopping cart!) The two most important forces in everyday life are gravity and friction. Gravity gives carts weight, and makes them roll downhill. Friction is produced when two surfaces in contact slide or rub. It is friction inside the cart's wheels that makes it eventually slow down and stop.

Newton's Laws

Isaac Newton published his Laws of Motion in 1687. The First Law explains how an object's motion does not change unless there is a force acting on it. The Second Law considers how the motion of a heavier object requires more force to change it than the motion of a lighter one. The Third Law states that forces come in pairs. So when a force is applied on an object, the object produces an equal force in the opposite direction. This means that the Earth's force of gravity pulls on you—and your own gravity pulls back on the Earth, with an equal but opposite force.

ISAAC NEWTON (1642-1727)

English mathematician and physicist Isaac Newton is one of the best-known scientists of all time. He described his Laws of Motion in 1687, in Latin, in his book "Philosophiae naturalis principia mathematica" (The Mathematical Principles of Natural Philosophy). As well as working out these laws, Newton designed an important type of telescope. He was the first person to realize that white light is made of light rays which are all the colors of the rainbow. He suggested that gravity was a universal force, possessed by all forms of matter, from atoms to planets. Newton's ideas about gravity may have been inspired in about 1666 by an apple falling from a tree in his garden at Woolsthorpe Manor, Lincolnshire.

Pressure

A person makes more of a dent in a soft floor wearing narrow heels than wide ones. The same force—the person's weight—pushes down on the floor in each case. But it is more "concentrated" with narrow heels, because these mak contact over a smaller area compared to wide heels. A scientist would say that the narrow heel exerts more pressure than the flat heel. Pressure is a measure of how much force is exerted on a certain area, usually measured in square feet or square meters. Not only solids exert pressure—liquids and gases do, too. The pressure of air inside an inflated balloon pushes outward in every direction If it did not, the balloon would deflate as the air was squashed by the force produced by the stretched rubber.

What is weight?

What we call "weight," and we measure in pounds or kilograms, is actually a force. It i due to the pull of gravity. Being a force, it should be measured in the units called newton

The weightlifter pushes upward on the bar, wit a force equal to gravity pulling down on the ba

The floor pushes upward on the weightlifter and weights, with a for equal to gravity pulling them both downward

Mr Universe

Gravity is a force of attraction (pulling things together) that acts everywhere in the Universe. It holds stars and galaxies together. It allows the Moon and artificial satellites to orbit the Earth. It also keeps us from falling off the Earth and flying away into space. This weightlifter has to work hard against the pull of gravity to li and hold up the weights on the bar.

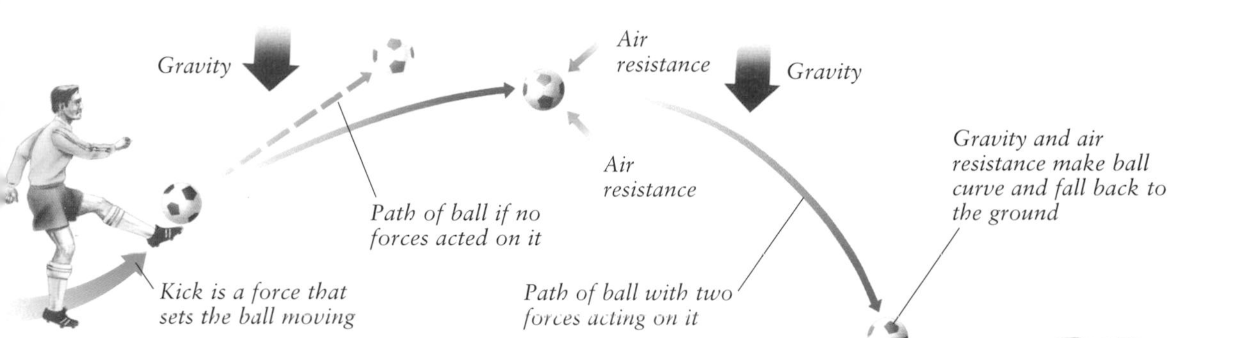

Something in the air

Kicking a ball shows Newton's Laws of Motion in action. The kick is a force that gives kinetic or moving energy to the ball. This overcomes its inertia, or tendency to resist movement. The ball sets off through the air. It would carry on for ever in a straight line, but two forces act on it. One is gravity pulling down. This slows the ball's ascent, makes it change direction, and accelerates it downward. The horizontal motion of the ball is unaffected by gravity, which can only pull downward. But another force acts, which is air resistance. This is caused as the ball pushes its way through air, knocking into the molecules. Air resistance makes the ball slow down. So the path of the ball through the air is a curve.

Full of air

Inside these balloons are countless millions of gas molecules, which make up air. They are dashing around at high speed. When these molecules hit the inside surface of the balloon, they produce pressure. This pressure keeps the balloon inflated. The rubber of the balloon itself pushes back with an equal but opposite pressure.

Air compressed or squeezed inside balloon

No force

There is no overall force on this parachutist. Gravity pulls him downward. But air resistance, caused by the parachute pushing past air molecules, acts in the direction opposite to motion. This upward force of air resistance, also called drag, exactly balances the downward force of gravity. So the parachutist carries on falling at the same speed. A bigger parachute would produce greater air resistance, and so the speed of the fall would be slower.

Weight a minute

The gravity of the Earth pulls this apple closer to the Earth's center. The gravity of the apple pulls back on the Earth with an equal but opposite force. We can see and feel the resulting balance of forces as the weight of the apple. The more mass an apple has—that is, the more atoms or matter it contains—the greater its weight.

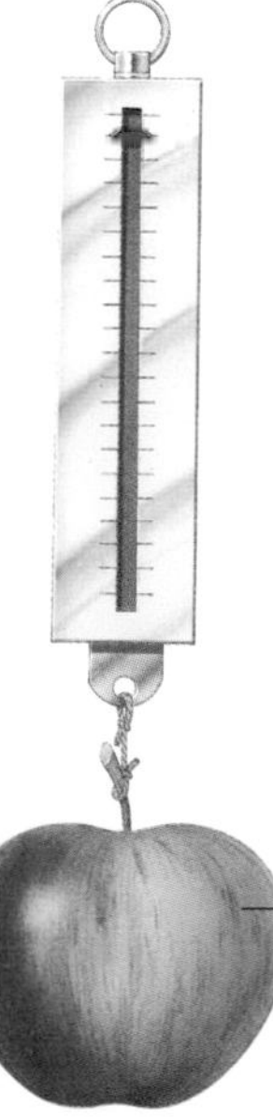

Apple and Earth attract and pull each other with equal forces

FLOATERS AND SINKERS

The pressure of water around an object, such as a ball, produces an upward force called upthrust (green arrow). The weight of the object is a downward force (yellow arrow). Heavier objects have a greater downward force. If this downward force is less than the upthrust, the object floats.

MACHINES

WHAT WE THINK OF as machines—bicycles and washing machines, for example—normally consist of many simpler machines working together. Scientists use the term "simple machines" to refer to levers, pulleys, wheels, screws, gears, and inclined planes (slopes). Simple machines make our work easier, usually by "magnifying" forces. For instance, a seesaw is a lever. If a person sits near the pivot of the seesaw, you can lift them up by pushing down on the opposite end. (A lever's pivot is its fulcrum.) It is harder to lift the person by pushing down nearer the middle. It is even harder lifting the person straight up, without using a seesaw at all! Even a simple slope (inclined plane), such as a ramp for a wheelbarrow, is considered to be a machine. You can raise the load in the wheelbarrow by simply pushing it up a long gentle slope. To lift the barrow and load directly, by the same height, would need a greater force. A corkscrew is another example of a device that uses simple machines. The screw thread twists into the cork as you turn it, while levers and gears make it easy to pull the cork out of the bottle.

Force and distance

You can reach the top of a hill by walking straight up its steep slope, or by walking up a less steep path that winds to and fro. The gentle slope is easier—but you have to walk farther. This link between force and distance is found in all simple machines. For example, a crane uses a combination of cables and pulleys called a block and tackle. The crane's engine can lift a very heavy weight by pulling on the cable with a relatively small force. However, the cable must move farther than the heavy object itself, because it passes up and down over several pulleys. The result is that the same amount of work is done in both cases. Lifting with block and tackle is easier than lifting without, but takes longer. The same applied to levers, such as using a screwdriver to prise open a can of paint (see right). You move the screwdriver handle easily over a longer distance. The tip pushes the lid over a shorter distance, but with greater force

Arms work as levers to pull out cork

Gears called rack and pinion lift the cork

Screw thread pushes down into cork

What a combination

The corkscrew is a clever combination of several simple machines. Imagine how it would work if its arms (levers) were shorter.

A complicated machine

The main arms of the backhoe each consist of three levers, linked at pivots. Each lever is pushed by devices called hydraulic rams, which are worked by fluid under high pressure. The backhoe's steering wheel is another simpl machine, the wheel and axle, a shown on the right.

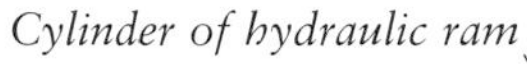

Cylinder of hydraulic ram

Piston rod of hydraulic ram

Pivot

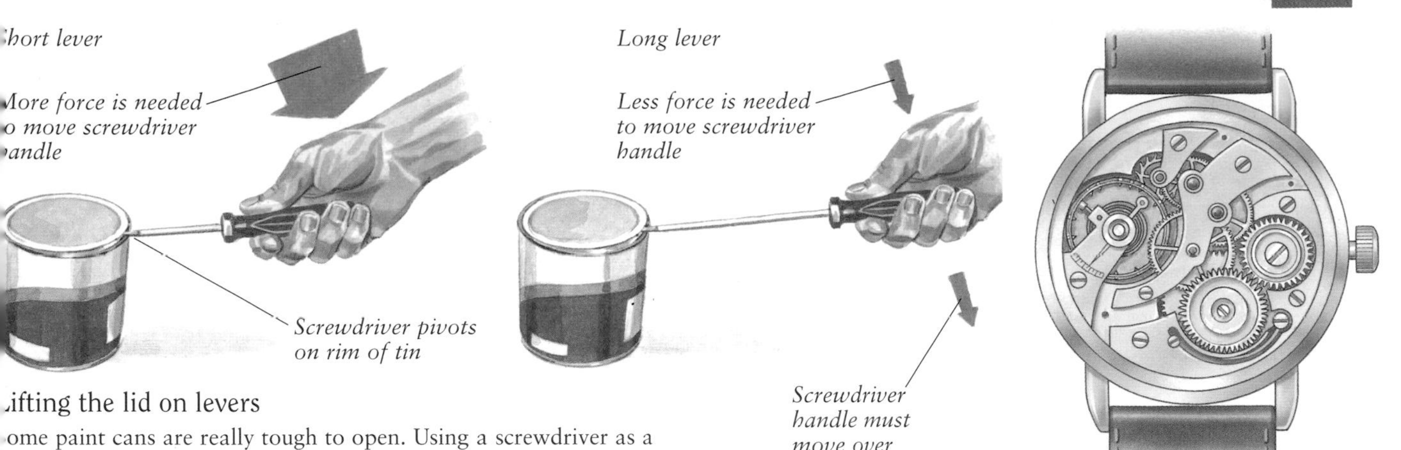

ifting the lid on levers

ome paint cans are really tough to open. Using a screwdriver as a ever makes it much easier. The longer the screwdriver, the less force ou need to apply to the handle—but the farther you have to push it.

Slow down

Machines can make movements slower and more controlled. Inside this watch, shown with its back plate removed, an uncoiling spring turns the gears that turn the hands. Without gears, the hands would speed around the face, and the clock would not keep good time.

You-turn

Some large vehicles, such as trucks and buses, have large steering wheels. They are easier to turn than a small steering wheel. But you have to spin them around a long way, perhaps several times, to turn the road wheels only a small distance. Steering wheels are examples of a type of simple machine called the wheel and axle.

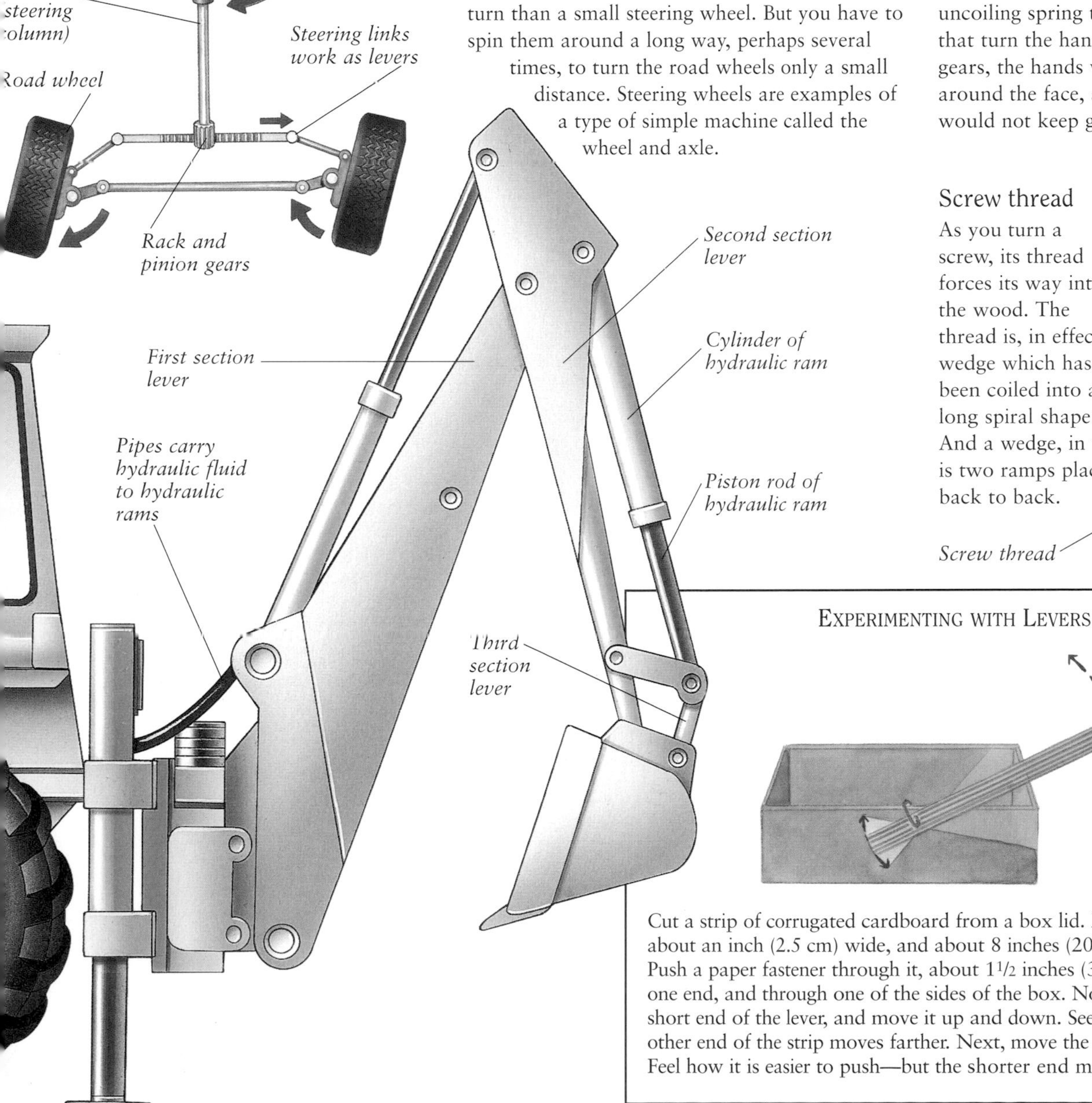

Screw thread

As you turn a screw, its thread forces its way into the wood. The thread is, in effect, a wedge which has been coiled into a long spiral shape. And a wedge, in turn, is two ramps placed back to back.

Screw thread

EXPERIMENTING WITH LEVERS

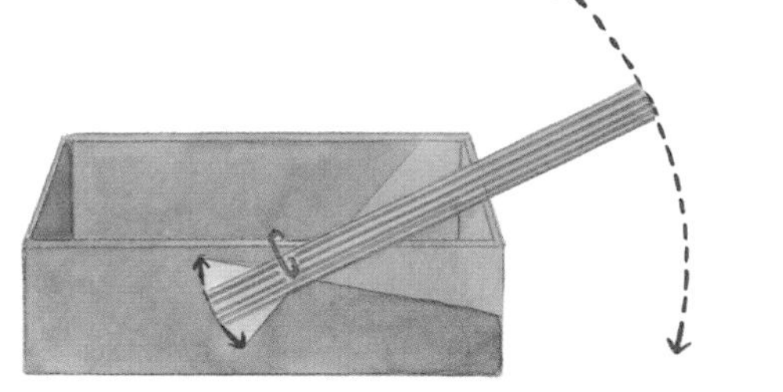

Cut a strip of corrugated cardboard from a box lid. Make it about an inch (2.5 cm) wide, and about 8 inches (20 cm) long. Push a paper fastener through it, about 1½ inches (3.5 cm) from one end, and through one of the sides of the box. Now, hold the short end of the lever, and move it up and down. See how the other end of the strip moves farther. Next, move the longer end. Feel how it is easier to push—but the shorter end moves less.

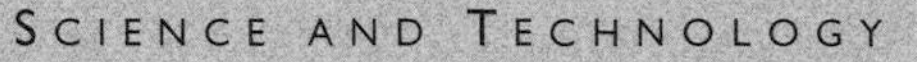

ELECTRICITY AND MAGNETISM

WHEN YOU SWITCH ON A FLASHLIGHT, billions of tiny particles called electrons move from the negative terminal of the battery, through the metal wires and the bulb, and back to the positive terminal of the battery—a complete circuit. This is an electric current since each electron, which is part of an atom, carries a negative electric charge, and movement of electric charges makes up a current. We use these moving electrons to do many useful things, including lighting and heating our homes, and powering televisions, music systems, computers, and hundreds of other electrical appliances that modern life depends on.
Static electricity, on the other hand, is caused by electric charges remaining in a fixed position. When you rub a balloon in your hair, electrons transfer from hair to balloon, and stay there. This leaves the hair positively charged, and the balloon negatively charged. The hair clings to the balloon because positive and negative charges attract each other. Two objects with the same charge—two rubbed balloons, for example—repel each other (push apart). Magnets also attract and repel each other, like electric charges. In fact, electricity and magnetism are very closely linked.

Magnets

All magnets have two ends, or poles. One end is called the north-seeking pole, or simply the north pole, and the other is called the south-seeking or south pole. A magnet's north pole always points north if a magnet is free to turn. When two north poles or two south poles are close together, they push apart. When a north pole and a south pole are brought close, they attract each other. A steel paper clip brought close to the north pole of a magnet becomes a magnet, with its south pole nearest to the magnet. This is how paper clips, and other objects made of iron or steel, cling to magnets.

Switch in closed or on position

Connection to positive batter terminal

Filament (thin coiled wire) inside bulb gets hot and glows as electricity passes through it

Connecting wires

Plastic insulating cover stops electricity leaking from wire

Connection to negative batter terminal

Make light work

Electrons moving around a circuit make an electric current. Their energy can make things happen. For example, when electrons flow through the filament of a bulb, they heat it up and cause it to glow. A switch can turn electric current on and off. If it is opened or off, the circuit is broken, and the electric current can no longer flow.

Electric charges

Electric charges are carried by electrons, which are particles found in every atom. All atoms have a central nucleus, which carries positive electric charge, surrounded by the negatively charged electrons. Within any one atom, there is normally the same amount of positive and negative charges, so the atom has no overall charge. It is said to be neutral. If electrons are lost or gained, then the balance of charge is upset, and the atom becomes charged. Inside a flashlight battery, electrons are separated from their atoms by chemical reactions. The electrons emerge at the negative terminal of the battery and move around a circuit. They are pushed by the negative charges of other electrons there, and pulled around the circuit by positive charge at the other terminal.

Back and forth

The electric current that is supplied to our homes, to power appliances such as hair-driers, is called alternating current, AC. This is because the electrons inside the cables move back and forth many times every second. The current supplied by a battery is direct current, DC, because the electrons move in only one direction (from the negative to the positive terminal).

Lines of force

You can imagine the invisible magnetism or magnetic field around a magnet, as lines of magnetic force. The closer the lines, the stronger the magnetism. The lines are closest, and so the magnetism is strongest, at the two poles of the magnet.

Electricity cables to electromagnet

On and off

Ordinary or permanent magnets exert their magnetism all the time. An electromagnet only does so while electricity flows, so it can be switched on and off. Electromagnets have hundreds of uses, from automatic door locks to cranes in car scrapyards.

Electromagnets

When an electric current flows through a wire, it produces a magnetic field around the wire. The magnetic field can be made stronger by winding the wire into a coil, the solenoid, and putting an iron bar in it. This is an electromagnet.

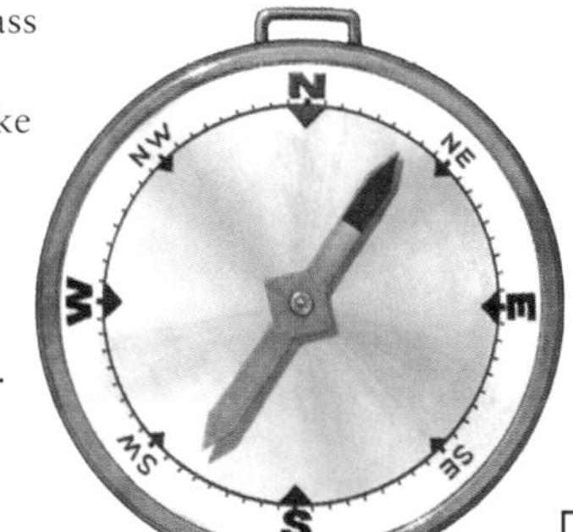

What a lineup

There is a good reason why a compass needle lines up north-south. Electric currents deep in the Earth's core make the Earth into a huge electromagnet. The compass needle is attracted to the poles of the Earth-magnet, which are found near the Earth's geographical North and South Poles.

Hair-raising

Static electricity can be produced by rubbing or friction. If a person becomes charged with static, each strand of hair has the same charge on its surface as all the other strands. Like charges repel, and so all the strands of hair repel each other and the main body. This makes the strands of hair stand up and "fly away." Gradually the charge leaks away into the air. This happens faster if the air is damp. Walking on a nylon carpet can also charge the body. When you touch a metal door handle, the charge leaps to it, with a small spark and shock.

EXPERIMENT: MAGNETIC FORCES

Find any two magnets, and investigate the forces between them. You can find out which end of each magnet is the north pole by hanging the magnet from a cotton thread. Use a compass to find out which way is north, and mark the north pole of each magnet in the same way. Then see what happens when you bring different combinations of poles together. What happens, for example, when you put a north pole next to another north pole? Or two south poles together? Does the pushing (repulsion) or pulling (attraction) force become stronger as the magnets move nearer each other?

LIGHT AND COLOR

INSIDE A TOASTER there is a wire called an element. When electric current passes through it, the element starts to glow with an orange light. If the toaster's element became hotter still, it would glow first yellow, then white like the filament of a light bulb. Light that is produced by hot objects is called incandescent light, and includes light created at the surface of the Sun. Sunlight is a jumbled mixture of many different colors of light, and you can see these colors in a rainbow. Light travels very fast, in straight lines, through many materials, including air, glass, and water. It bends or refracts as it passes from one material to another. This is useful in making lenses, which bend light in just the right way to produce eyeglasses, telescopes, and microscopes. When light cannot pass through a material, it may be absorbed, or it may be reflected or bounce back. Colored surfaces absorb some colors while reflecting others. A red sweater, for example, reflects only red light.

Reflection and refraction

Reflection is light bouncing off an object. Refraction is light bending as it passes from one material to another. Both are very useful. For example, if light did not reflect from objects that do not produce their own light, we would only be able to see light sources. To see a book in a dark room, for example, we must position a lamp so that light falls onto the book and reflects off it into our eyes. Mirrors and other shiny surfaces reflect light uniformly, and this means that you can see images in them, such as your face, for example. You can observe refraction if you look at an object under water. Light from the object bends as it passes from the water to the air. We make use of refraction by shaping glass into lenses that can produce images by focusing light.

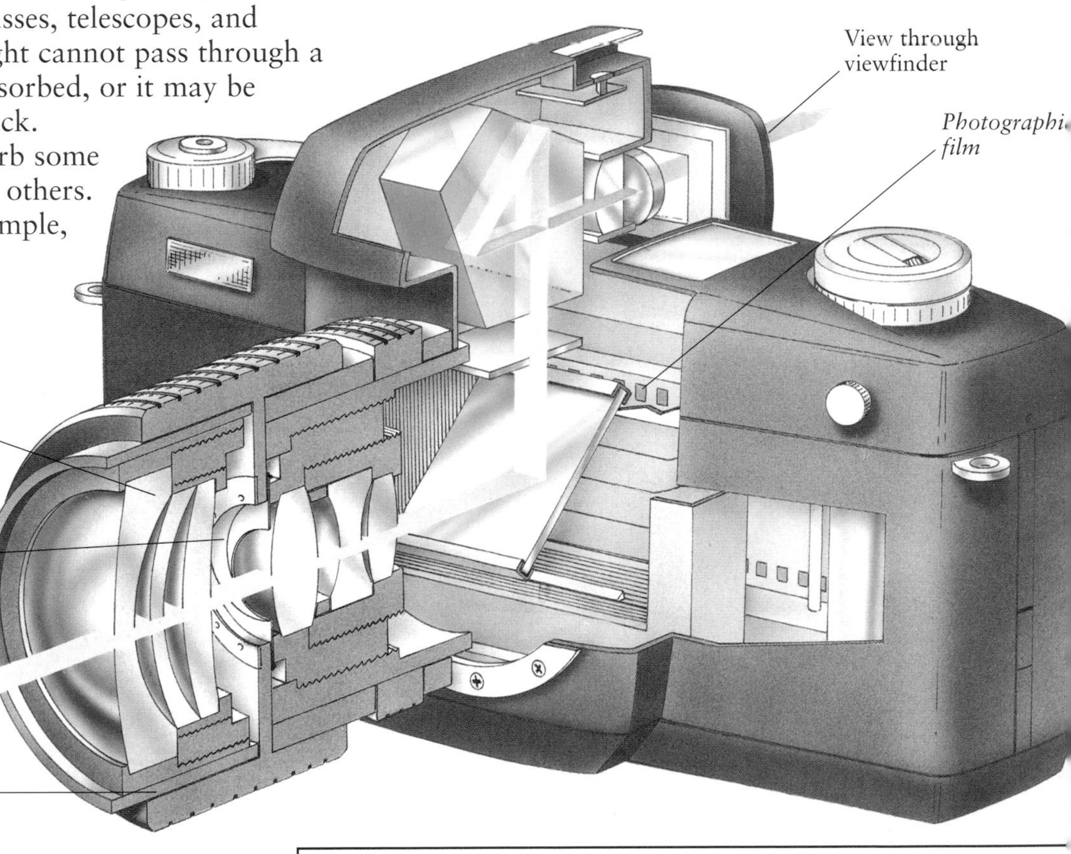

View through viewfinder

Photographic film

Lens system of multiple lenses

Diaphragm controls amount of light entering camera

Path of light rays

Screw mechanism for focusing

What a picture

At the back of a camera there is a flat piece of photographic film. Light passes through the lens at the front, and reaches the film so that light from any one point of the scene in front of the camera falls on the same point on the film inside it. This is called focusing. The film records the pattern of light making up the image, by undergoing chemical changes where light hits it. (In a digital camera, the photographic film is replaced by light-sensitive electronic panel called a CCD, or charge coupled device.)

FACTS ABOUT LIGHT

- It takes only a hundred-millionth of a second for light to travel 3 feet.
- The spectrum does not consist of seven distinct colors, as many people believe. It is a band of colors that varies gradually from red through to purple.
- There are three different types of color-sensitive cell in the eye. People who are color-blind have one type of cell that does not work properly.
- The surface of the Sun is at a temperature of about 10,800°F (6,000°C). At this temperature, it gives off white light. As sunlight passes through Earth's atmosphere, air molecules bounce some of the blue light in all directions, which is why the sky is blue.
- Lenses that bulge out in the middle are said to be convex. Magnifying glasses are convex lenses.

A colorful spread

The colors of the rainbow are all present in white ight. They separate as they pass through a riangular piece of glass or plastic, called a prism, ecause they all bend by different amounts. This pread of color is called the light spectrum. Millions of raindrops in the sky do the same job, which is why you only see rainbows when the Sun shines on a rainy day. The colors are known by tradition as red, orange, yellow, green, blue, ndigo, and violet.

Prism

White light enters prism

Colors of light spectrum

Upside down

When you look through a lens, objects beyond the lens may look larger (magnified) or smaller (diminished) and upside down. This is because the lens refracts or bends the light rays passing through it.

Concave lens diminishes image

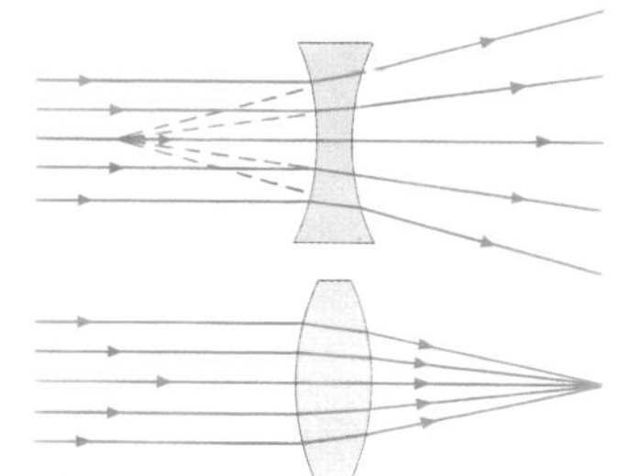

Convex lens magnifies image

On reflection ...

Light that falls on your face reflects off it in all directions. If some of the light hits a nearby mirror, it will reflect uniformly. Some of that reflected light then enters your eyes, and you see an image of your own face. Flat or plane mirrors give undistorted images. The mirror in this picture is curved, and so it produces a wobbly or distorted image.

Watching television

Our eyes contain three types of color-sensitive cell: red, green, and blue. The tiny dots or phosphors on the inside surface of a television screen come in the same three varieties: red, green, and blue. By mixing light from these phosphors in different combinations, it is possible to fool the eye into seeing any color.

Electrons from red, green, and blue guns

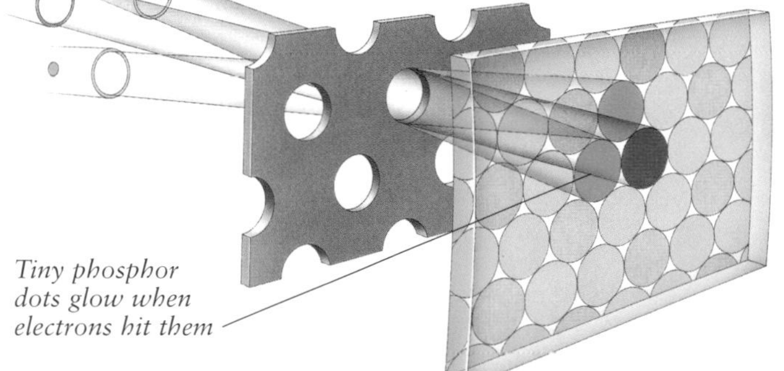

Tiny phosphor dots glow when electrons hit them

Glow in the dark

ncandescence is not the only way of producing light. There are dyes on the inside surface of a television screen called phosphors, which give out light without being hot. They glow by a process called luminescence. Other luminescent sources of light include glow-in-the-dark signs and some ypes of fluorescent lamps. The light is produced by electrons inside the atoms that make up a luminescent material. Electrons that gain energy, by absorbing light or electrical energy (but not heat), for example, give out light as hey lose that energy. In a glow-in-the-dark sign, electrons gain energy from light that falls on them during the day. During the night, the billions of electrons give out their extra energy as an eerie green glow of light.

LIMITED REFLECTION

Hold a brightly colored object, such as a red sweater, near a white wall in a dark room. Shine a flashlight or lamp onto the object, and look at the wall. You should see a patch of light illuminating the wall. What color is the light on the wall? With a red sweater, it should be red. Only certain colors of light reflect off colored objects. White objects (such as the wall) reflect all colors. This is why you can see the patch of light reflected from a colored object, whatever its color. With a matt black object, you should get no reflection, because black objects do not reflect any light.

Sound

IF YOU TOUCH THE SIDES of your neck as you speak, you will feel small movements or vibrations. Inside your throat, there are two flaps of skin called vocal cords. These move to and fro (vibrate) thousands of times every second when you speak. Plucked guitar strings produce sounds when they vibrate, too. In fact, all sound is caused by vibrations. A vibrating object disturbs the air around it, sending waves of sound out in all directions, like ripples on a pond. The vibrations pass through the air, and those that enter our ears, we hear as sound. The vibrations of sound also pass through liquids such as water (you can hear under water) and through solids (you can hear sound through walls). How loud or quiet a sound is depends on how far away you are from the source of the sound, and how much the sound source moves to and fro as it vibrates. Sound may also be high- or low-pitched. A rapidly vibrating object produces a higher-pitched sound than an object vibrating more slowly.

Sound waves

Vibrating objects, like drums, disturb the air in a similar way to a finger moving to and fro at the surface of a still pond. Sound waves travel out in all directions. The bigger the vibrations, the taller the waves, which means louder sounds. The height of a wave is called its amplitude. The faster the vibrations, the higher the sounds in pitch. The speed of vibration is called the frequency. Gradually the energy of sound waves fades, which is why things sound louder when you are closer to them

Cymbal vibrates fast and produces high-pitched (high frequency) sound

Drum vibrates slowly and produces low-pitched (low frequency) sound

Bass drum vibrates ve slowly and produces very low-pitched (low frequency) sound

Vibrations

Vibrations of sound can be recorded using microphones, which make copies of the vibrations as varying electric currents. When the sound waves reach a microphone, they cause a tiny magnet to move to and fro within a coil of wire, inside the microphone. The electrical signals are produced inside this coil (see opposite). The signals from a microphone can be made larger by an electronic device called an amplifier. The output of an amplifier is a bigger, more powerful version of the varying electric currents. These make a cone inside a loudspeaker vibrate in exactly the same way as the original source of sound—but much louder. The output from the microphone can also be recorded as tiny patterns of magnetic patches on magnetic tape. There are many other ways of recording sounds, such as on compact discs (CDs). But each one involves making a copy of the original electrical signals, like those from a microphone.

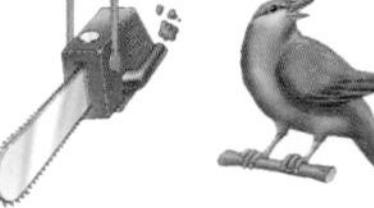

Sounds good

The scientific study of sound is known as acoustics. Experts called acoustic engineers help to design and build recording studios, theaters, and other sound and music venues. They predict how sound waves will travel through the space, and how they will be absorbed or reflected by the walls, ceiling, floor—and people. They advise on materials to direct sound clearly to the audience.

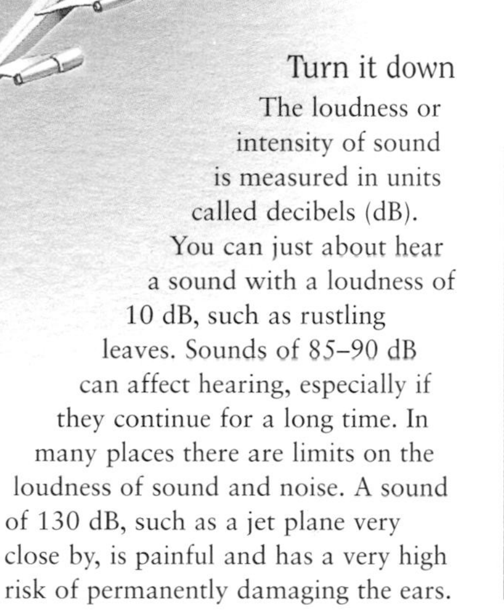

Turn it down

The loudness or intensity of sound is measured in units called decibels (dB). You can just about hear a sound with a loudness of 10 dB, such as rustling leaves. Sounds of 85–90 dB can affect hearing, especially if they continue for a long time. In many places there are limits on the loudness of sound and noise. A sound of 130 dB, such as a jet plane very close by, is painful and has a very high risk of permanently damaging the ears.

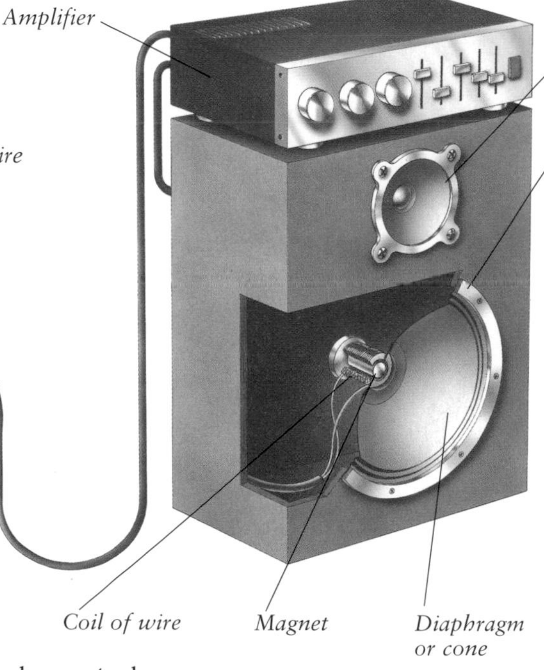

ood vibrations

side most microphones is a in piece of flexible material lled a diaphragm, which brates to and fro as sound aves hit it, a bit like a cork obbing up and down with the pples in a pond. The aphragm is fixed to a coil of ire which is near a magnet. he movements of the coil in e magnetic field produce ectric currents in the wire.

Under control

The cone or diaphragm of a loudspeaker works in the opposite way to a microphone. It produces sound as it is made to vibrate by powerful electrical signals from the amplifier.

RUBBER BAND GUITAR

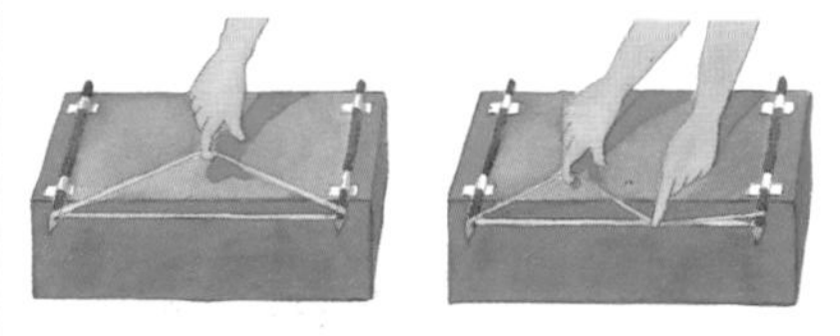

Tape two pencils across an upturned shoe box, one near each end. Stretch a thin rubber band over the pencils and pluck the rubber band. The rubber band vibrates and disturbs the air around it. The vibrations are also passed on to the box, and this makes the sound louder than it would be without the box. Now press a finger about halfway along the rubber band. Now, the rubber band vibrates more times every second when you pluck it, and the note is higher.

ELECTRONICS

COMPUTERS, HI-FIS, TELEPHONES, and even some washing machines and toasters use electronics—the careful control of often tiny electric currents. Inside a computer, these tiny electric currents represent numbers, letters, and commands. They move around complicated electronic circuits which consist of components—small devices that affect the currents in different ways in various parts of a circuit. A component called a resistor is used to reduce the current flowing through its part of a circuit. A capacitor stores electric charge as current flows through it, ready for release at the correct time. A diode allows electric current to pass through in one direction only. Combining the various functions of these many different components, in complicated circuits, allows us to have a host of electronic devices for everyday use. Hundreds or thousands of components can now be made out of a single sliver or wafer of the material called silicon. The components are microscopic in size and are made ready-connected to each other, or integrated. Most modern electronic equipment includes these integrated circuits, also known as microchips or silicon chips.

Making the connection
Inside all modern electronic devices are circuit boards like the one on the right. These boards hold electronic components, which are connected together by metal tracks to form a complicated electric circuit. Most electronic devices contain integrated circuits.

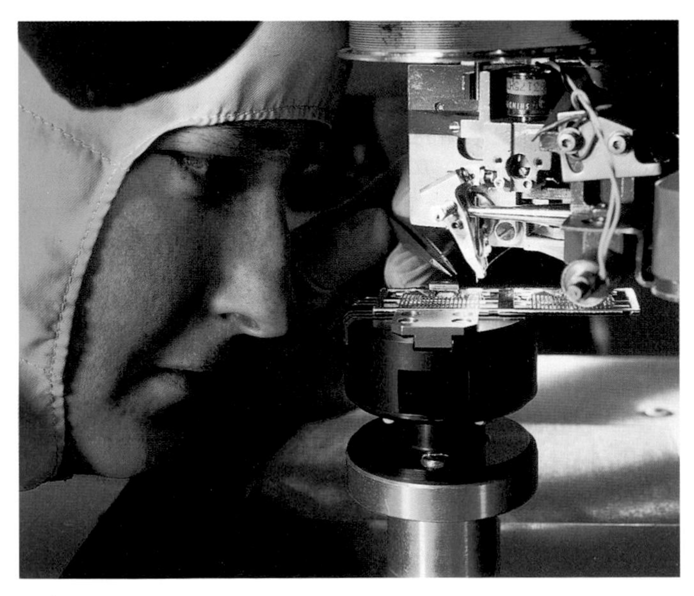

Making a circuit
The manufacture of integrated circuits is a sophisticated process, involving many stages of treatment to a slice of pure silicon crystal. During the process, the silicon is subjected to heat and attack by acid and ultraviolet light.

Semiconductors

Most metals allow electric currents to pass through them easily. They are said to be good conductors of electricity. Plastics, wood, glass, and many other materials do not conduct electricity very well at all. They are insulators. Some materials are in between—they conduct electricity fairly well, or only in some circumstances. This makes them very useful in electronic devices. For example, some semiconductors carry electricity if light falls on them, so they are used in light-sensitive equipment such as video cameras.

Diodes are made from two different types of semiconductor, as a combination that conducts current in only one direction. Transistors are also made of semiconductors (see opposite). They can be used to make an electric current larger. This is amplification, and is important in sound recording and playback. Transistors are also used as switches, to turn currents on and off. This is put to use in calculators and computers, which process numbers as electric currents that are either "on" or "off."

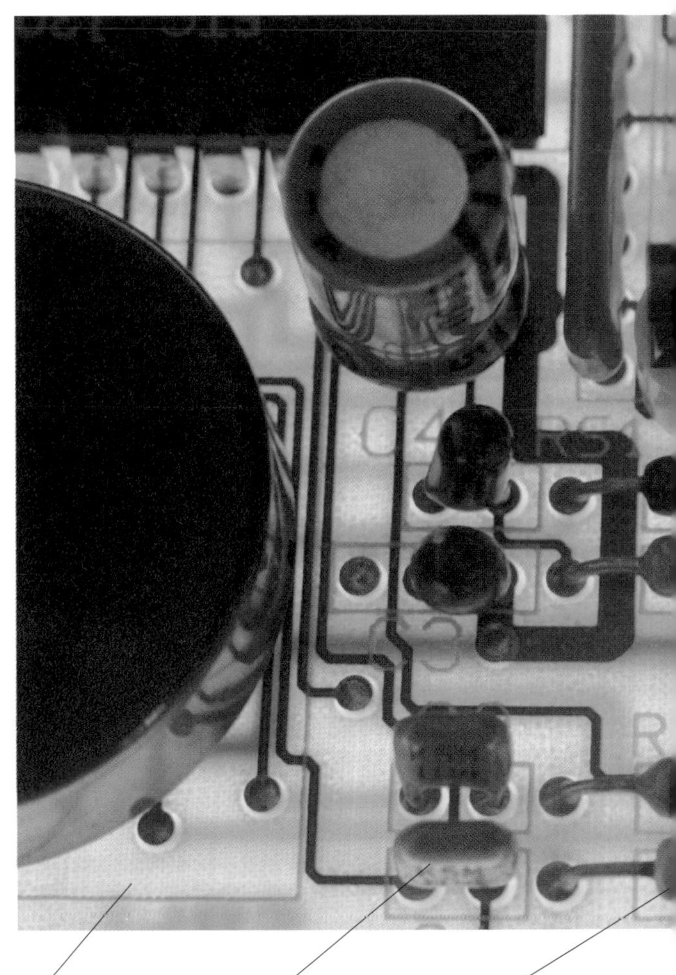

Circuit board is made from an insulating material such as plastic, ceramic or fiber-board

Capacitor stores electrical energy as electric charge, until it reaches a certain level or for a specific time, then releases it

Resistors resist the flow of electric current and reduce the amount passing through that part o the circuit

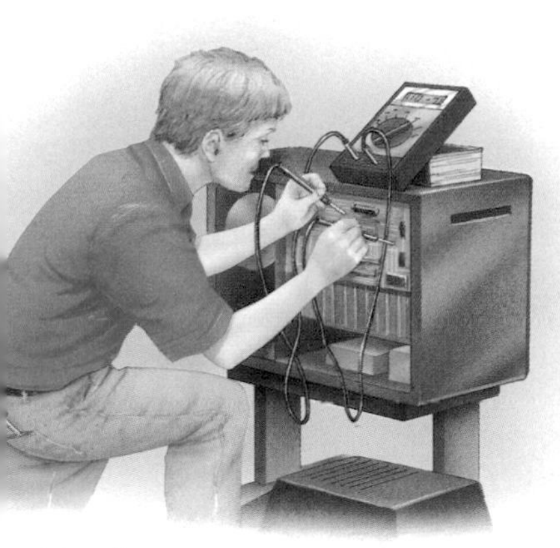

ust testing

A circuit may need to be tested, to make sure it is working properly. The probes of this multimeter touch different points in the circuit, and the meter displays how much current flows through that part of the circuit.

Low prices!

Since it has been possible to manufacture integrated circuits at low cost, the range and availability of electronic goods has increased dramatically. There are few houses in developed countries without a television and a telephone, for example.

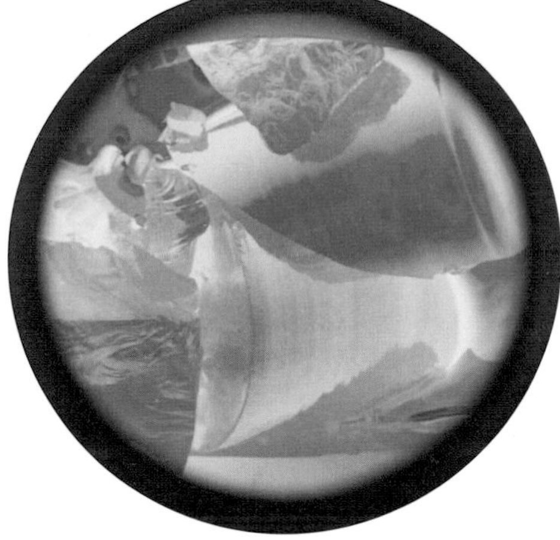

The chips are down

The best-known semiconductor is silicon, shown above as wafers of silicon crystal. It is the basis of many electronic components, including tiny integrated circuits. This is why integrated circuits are sometimes called silicon chips.

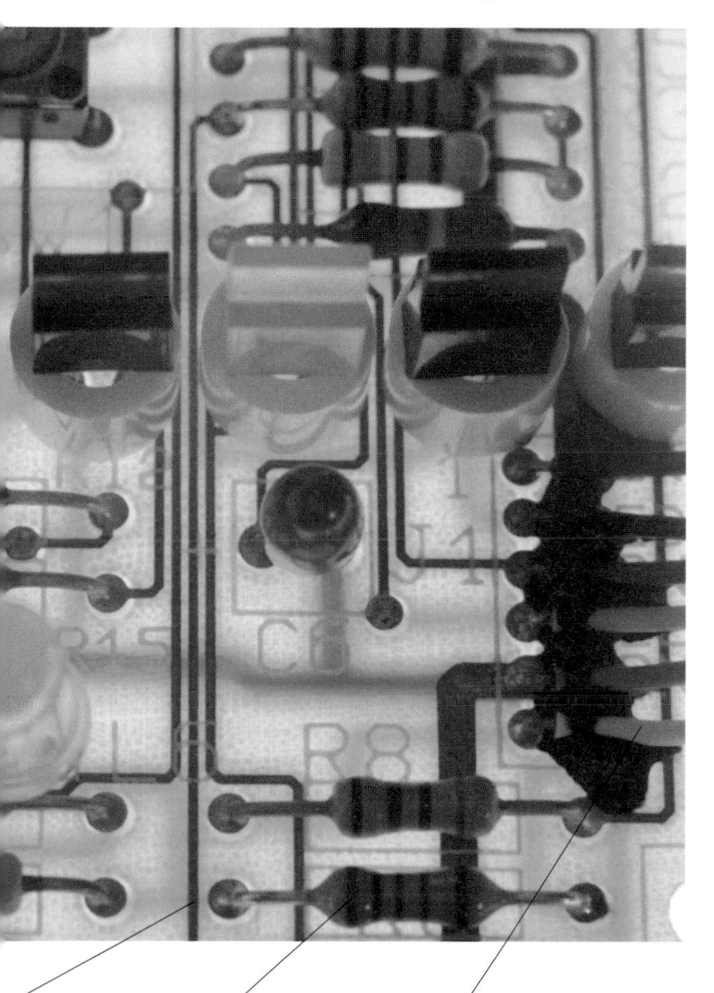

Metal tracks "printed" on the circuit board (PCB) act as wires to carry electricity to certain components

Resistors are color-coded, the order and color of the bands indicating their resistance in units called ohms

Solder connections link the circuit board via wires to other components such as the electricity supply

Variable electronic components

The metal "legs" of each component carry electric current to and from metal tracks or pins on the circuit board. A variable resistor, like an ordinary fixed resistor, controls the current flow through a circuit. But a variable resistor can be adjusted for higher or lower resistance. Volume knobs and similar controls on music systems are normally variable resistors. Turning the volume control up allows more current to flow to the loudspeakers, making the loudspeakers produce a louder sound. Variable capacitors are also adjustable and are often used for the tuning controls on radios. Transistors, like diodes and integrated circuits, are made of semiconductor materials such as silicon. They have three connections or wires. The current flowing into one controls the amount of current flowing between the other two. Transistors can work as switches, amplifiers, oscillators (to reverse the direction of electricity rapidly), and photocells to control current in response to light levels.

LED

As its name suggests, a light-emitting diode (LED) is a type of diode that gives out light. The light is of a very precise color which is determined by the materials from which the diode is made. LEDs are normally encased in a plastic of the same color, to protect them from damage. They are not usually as bright as light bulbs, but they use much less electricity. They are also extremely reliable, rarely burning out.

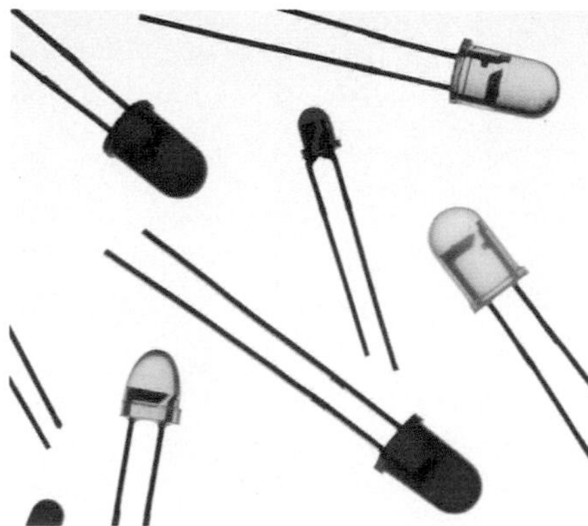

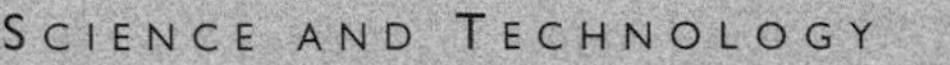

COMPUTERS

WHEN YOU TYPE WORDS using the keyboard of a personal computer (PC), pulses of electricity pass along wires into the main computer unit. Inside this is the central processing unit (CPU), which processes data (pieces of information) put into it. The data are in the form of "on" (1) and "off" (0) electrical pulses that represent letters, numbers and simple instructions. For example, the code for the letter w is 01110111. The 1s and 0s are called bits, short for binary digits. A group of eight bits is a byte. One kilobyte (kB) is about one thousand bytes, one megabyte (MB) is about one million bytes, and one gigabyte (GB) is one billion.

As well as signals from the keyboard, mouse, scanner, camera, or other input device, the CPU receives data or information from the computer's memory. There are two main types of memory inside the computer, RAM and ROM (see right). In addition, data and programs are stored on disks, including the main or hard disk.

Text, numbers, pictures and sounds can all be "digitized," represented by bits. A high-quality sound that lasts one second, for example, can be represented by about 700,000 bits (90 kB). The results of processes inside the CPU are sent to output devices such as the screen, printer or loudspeaker.

ALAN TURING (1912-1954)

Mathematician Alan Turing was very important in the early development of computers. He studied maths in England and the United States. During World War II he developed a computer called ENIGMA, which could decode secret enemy messages. Turing was one of the first people to realize how computers might one day be truly intelligent. He suggested that artificial or machine intelligence would be recognized when a person communicated with the machine and asked it questions over a remote link (like the Internet), and assumed it was a person, not a machine.

How computers remember

There are many different devices for storing computer data. Some are integrated circuits (microchips) which have arrays of millions of electronic components called capacitors. Each capacitor can hold either a tiny electric charge (bit 1) or no charge (bit 0). RAM, random access memory, is this type. It holds a temporary store of sets of instructions called programs, along with the results of calculations carried out by the CPU. ROM, read-only memory, is similar, but its data cannot be changed or overwritten. ROM holds the basic system programs that the computer needs to start, or boot up. A CD-ROM is very different. It is a compact disc (CD) that holds huge amounts of data. This may be programs, text, pictures, animations, and sounds, as for an encyclopedia. A similar disk, DVD or digital versatile disk, holds even more than a CD-ROM.

What's inside?

The main unit of a PC contains various microchips, including the CPU, and several RAM and ROM chips. All the parts within the unit are attached to the main circuit board or motherboard and connected by metal tracks called buses or ribbon connectors.

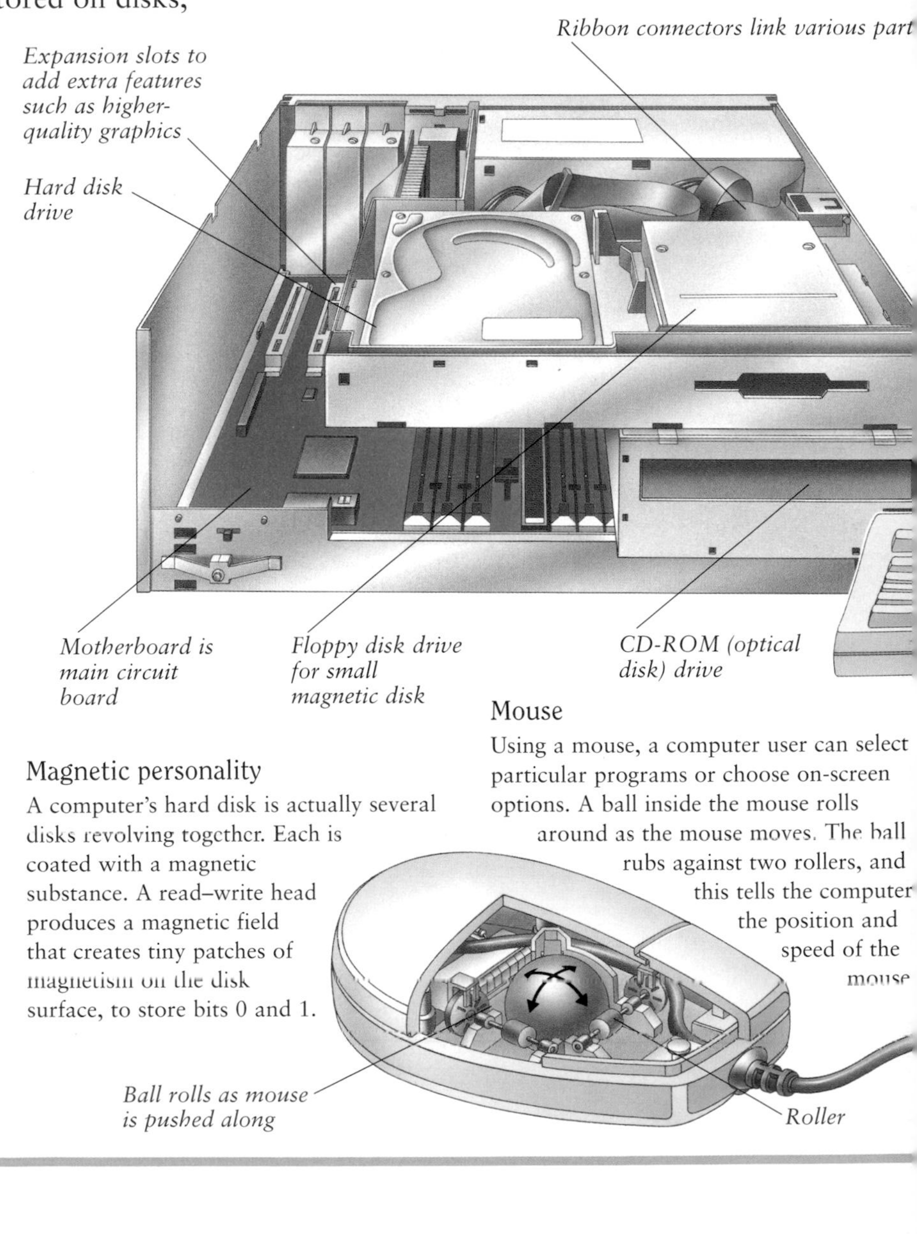

Magnetic personality

A computer's hard disk is actually several disks revolving together. Each is coated with a magnetic substance. A read–write head produces a magnetic field that creates tiny patches of magnetism on the disk surface, to store bits 0 and 1.

Mouse

Using a mouse, a computer user can select particular programs or choose on-screen options. A ball inside the mouse rolls around as the mouse moves. The ball rubs against two rollers, and this tells the computer the position and speed of the mouse.

creen

screen, or monitor, allows a computer user to view the processes id results happening inside the machine, and also to interact with e computer. When the user moves the mouse (see below left), a ointer moves across the monitor screen. The words or numbers ped on the keyboard appear on the monitor, too. The screen can splay pictures, diagrams, movies, and animations as well as ords and numbers. Monitors work in the same basic way as the reen of a television set. However, some screens, especially those 1 small laptop computers, have much flatter, lightweight screens hich are liquid crystal displays (LCDs).

On the desktop

The most familiar computers are desktop PCs (personal computers), consisting of a main unit with the hard disk, a monitor screen, a keyboard, a mouse, and loudspeakers. There is often a printer too, which can make hard copy—printed documents such as letters or pictures. There may also be a scanner, which works like a photocopier, and allows a drawing or photograph to be converted into computer data. Computer programs are called software, whereas hardware is the computer and other devices connected to it. Applications are programs that allow the computer to be used for a particular task.

eyboard

he main way of inputting words and numbers into a computer is / typing them on a keyboard. Each time a key is pressed, ectrical contact is made—so each key is like a switch. When a urticular switch is activated, a series of electrical pulses is sent to the computer. However, the time of the keyboard may be nited. In the future, we may well simply speak our instructions id data to computers. It is already possible to talk to some mputers, if they have microphones connected and they are aded with programs called voice-recognition software.

Each key is like a switch and makes electrical contact when pressed

The key sends a series of electrical pulses or bits when pressed

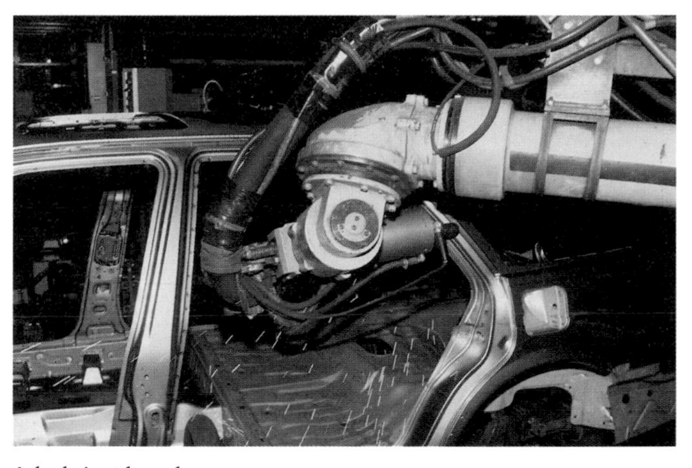

A helping hand

The data output by a computer can be used to control machines called robots. A computer-controlled robot can carry out the same task in precisely the same way, thousands of times daily. It can also carry out dangerous tasks. But these robots can also sometimes put people out of work.

On the move

Many people who work use computers, if only to write letters, and store names, addresses, and other information. For those who spend much time on the move, a lightweight portable or laptop computer like this one, powered by rechargeable batteries, is very useful.

ist worker

odern computers can handle ormous amounts of data very iickly. This means that they n be used to create and show imations or to edit and play deo and sound. The use of unds, animations, and videos well as text and images is lled multimedia.

Flow of information

The arrows on this diagram show the relationship between the CPU, RAM and ROM, and input and output devices such as the monitor, keyboard, and printer.

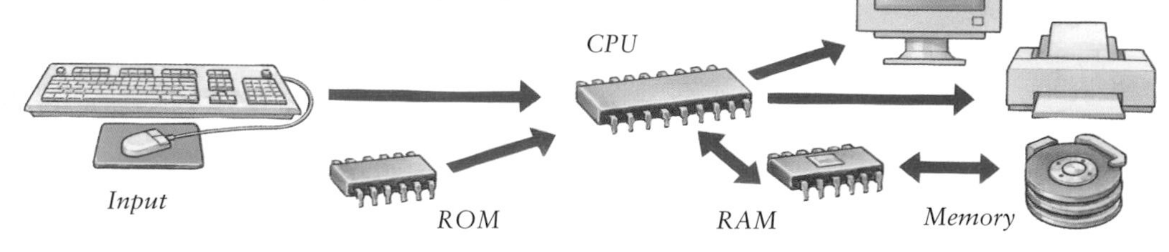

CARS

MOST CARS ARE POWERED by engines that run on gasoline, which is made from oil. Hundreds or thousands of times every second, tiny sparks inside each cylinder of a car's engine ignite a mixture of gasoline and air. The mixture explodes and that pushes a piston to the bottom of the cylinder. The piston is connected to a shaft, called the crankshaft, which turns as the piston is pushed down. Most cars have four or six cylinders, each with a piston. As the shaft turns, it brings pistons up to the top of the cylinders again, ready for the next explosions. The rotation of the shaft – powered by the explosions in the cylinders – turns gears in the transmission. The gears turn another shaft, which turns the car's wheels. Cars have batteries, which supply electric power for lights, and to produce the sparks to ignite the fuel in the piston. Some cars have diesel engines, in which there is no spark. The fuel and air mixture ignites when it is squashed by the piston rising up the cylinder.

The way out of the jam

The number of cars in the world has risen continuously over the past hundred years, and there are now so many cars in towns and cities that there are frequent traffic jams. Burning gasoline or diesel pollutes the air by releasing carbon dioxide gas and other substances. Most car designers attempt to make their cars more efficient, so that they drive just as far using less fuel. In the future, we may be driving solar-powered electric cars that receive their power from the Sun. As well as producing noise, traffic jams, and pollution, cars can be dangerous. Thousands of people are killed on the roads each year. Designers are always trying to find new ways of making cars safer for the driver, passengers, and pedestrians. Assisted brakes, seat belts, and airbags are all used to make cars safer.

The first car

The first really practical car to be produced was the three-wheeled Motorwagen. It ran on gasoline but had a top speed of just 12 miles per hour. The Motorwagen was designed and built by Karl Benz in 1885. Since then, more than 300 million cars have been built.

On track

Speed is the most important feature of racing cars. To achieve this, they must have powerful engines and streamlined bodies that push through the air with least resistance. Racing cars are normally close to the ground, which makes them easier to control as they speed round corners.

Inside the cylinders

The pictures below show the order of events inside a particular cylinder. There are four stages to the four-stroke cycle of a petrol engine, beginning with the intake (1) of the gasoline–air mixture. The mixture becomes compressed (2) as the piston moves upward, and then it is ignited (3) by a spark. As the piston comes around again it forces the exhaust gases out of the cylinder (4).

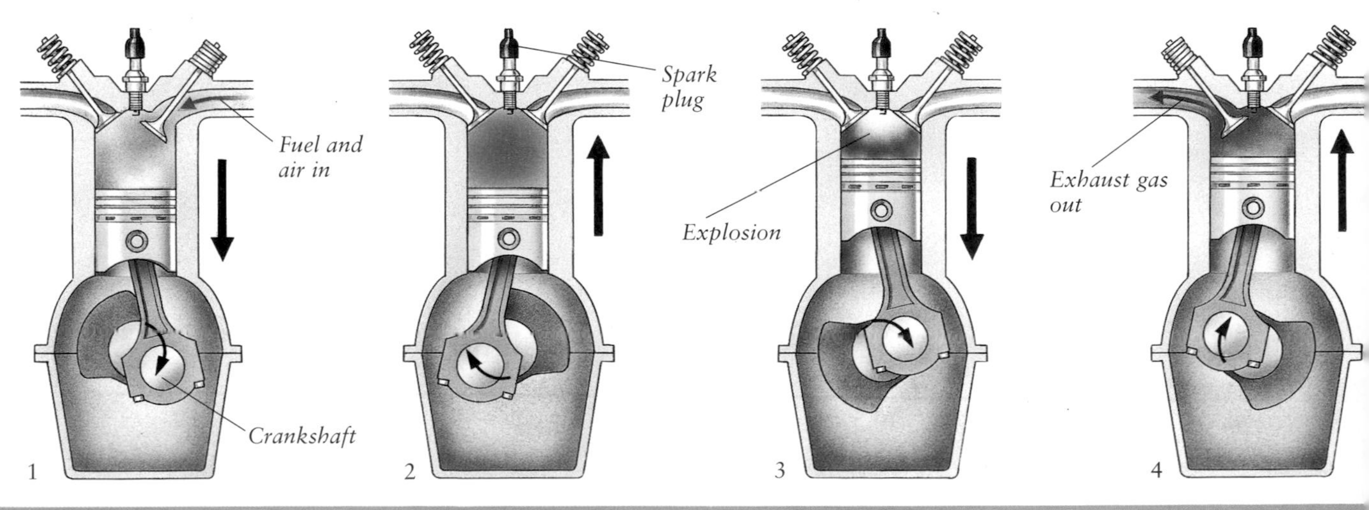

Sheer weight of traffic

Regular traffic reports on radio and television use the phrase "sheer weight of traffic" to explain why particular roads are at a standstill. Within towns and cities, public transportation, such as buses and trams is a very effective way of reducing traffic.

Testing the flow

As a car moves along the road, it has to push its way through the air. The smoother – or more aerodynamic – the shape of the car, the easier it is for it to push through the air, and the less fuel it uses. Below, a model of a car is being tested. A stream of mist passes over the body of the car in an artificial wind (as if the car were moving through still air at high speed). The less the car disturbs the mist stream, the more aerodynamic it is.

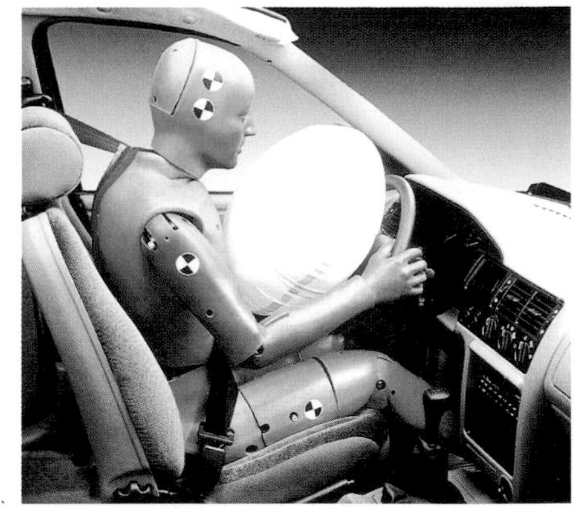

Cushioning the blow

Most modern cars are fitted with airbags. These are installed in the steering wheel or the dashboard, and they automatically fill with air if the car is involved in a crash. When the driver is thrown forwards, he or she hits the airbag, which cushions the blow. An airbag will only work in a head-on collision, and will not be effective in a side impact or a roll.

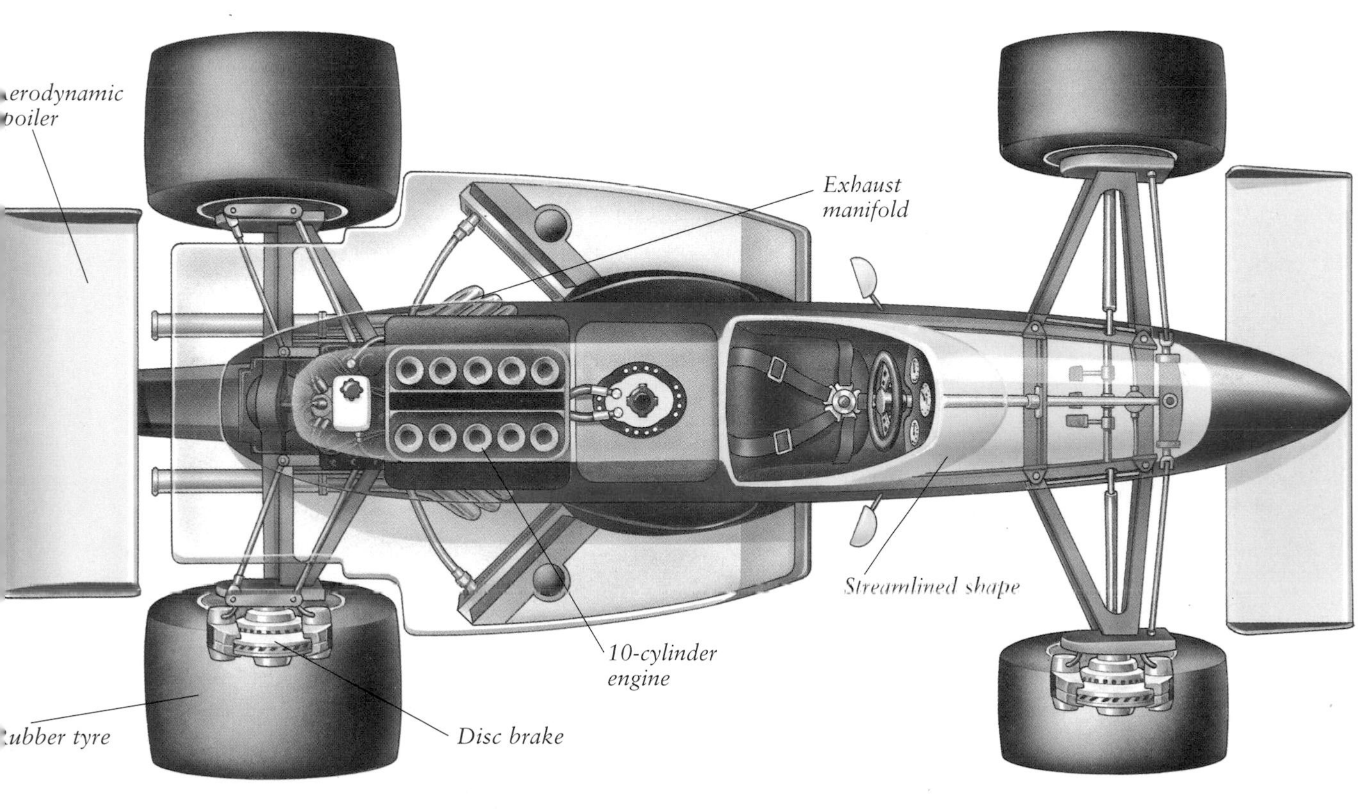

Driving force

As well as the engine, transmission and the shafts that turn the wheels, a car needs brakes to stop it, lights to see and be seen in the dark, a cooling system for the engine, and windshield wipers and demisters to allow the driver to see clearly. All of the different systems of a modern car work together to make driving as safe and easy as possible. Formula one racing cars like this one (above) do not have the same systems as a modern family car. They do not need headlights or demisters, for example. But they do need very powerful engines, so that they can move at more than 200 miles per hour.

KNOW YOUR CARS

- Some cars run on compressed air, which is carried in tanks in the car.
- The main problem with electric cars is that they do not have the same range as gasoline or diesel cars.
- Grand Prix motor racing began nearly 100 years ago, in 1906.

BOATS AND TRAINS

THERE ARE MANY TYPES of boats and ships. They carry people or cargo, or are used for sport and leisure. Boats stay afloat because of a force called upthrust, which is produced by the pressure of water pushing on the hull of the boat. To produce enough upthrust to float, the boat sits quite low in the water. It displaces, or pushes aside, huge amounts of water. Most boats have some source of power in order to move through water. This may be oars to push the boat along by human power, an engine-powered propeller under the water, or sails. In order to navigate, many boats have sophisticated systems controlled by computers. They have radar and sonar systems that use radio waves or sound waves to detect the depth of the water, and to avoid collisions with other craft, floating objects like icebergs, and rocks hidden just below the surface.

Hoist the sails
The wind pushes sails, and the sails are attached to a boat, and this is how a sailboat moves. The direction of movement depends on the direction of the wind, the angle of the sails, and the position of the boat's steering rudder under the water. By changing direction repeatedly in a zigzag maneuver called tacking, a boat can sail gradually into the wind.

Pushing through
Most boats have propellers, which are shaped like fans. Just as fans move air as they turn, a propeller pushes the water backward, forcing the boat forward.

Floating low in the water
A boat or ship floats lower in the water if it is heavily laden. A line on the side of a ship, called the Plimsoll line, shows what level the water should reach on the boat's hull, in various types of water. If the water level is higher than the Plimsoll line, the boat is too heavily laden and could sink.

From steam to electric

The first trains were powered by steam engines. Their invention changed travel and industry completely. Cargo could be transported by train and people could move around much faster than in the past. Most modern trains are electric or diesel-electric. Electric trains pick up electric power from an extra rail or an overhead cable. Diesel-electric trains produce their own electrical power, in generators that are driven by diesel engines. In each case, the electrical power drives electric motors that turn the train's wheels. Trains are more efficient than cars. They use less fuel per person over distance. Traveling from city to city on a train instead of a car means that there are fewer cars on the road, too.

Over the points

Unlike cars or buses, trains cannot steer in the direction they want to go. At certain places the tracks cross over, allowing trains a choice of destination. These are called points. In a modern railroad system the points are controlled by computers with safety backups.

The right shape

The shape of train wheels is very important. If they were flat, the wheels would slide off the tracks.

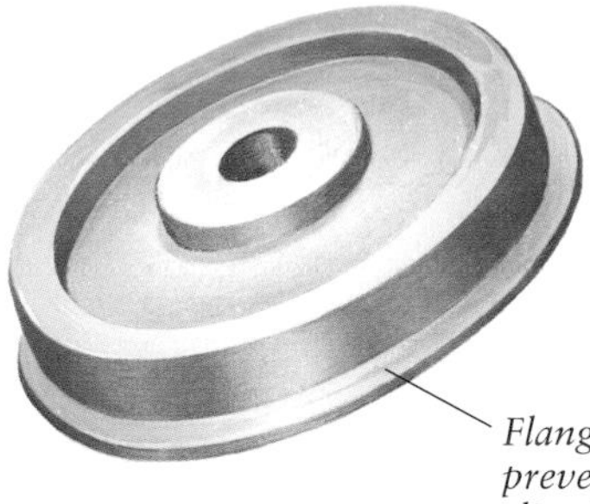

Flange on side prevents wheel slipping off rail

Under its own power

Most modern trains are powered by electric motors. The electric motor is very efficient, turning more than 90 percent of the energy fed into it, as electricity, into the energy of movement. A gasoline engine is only about half as efficient.

Air craft

Powerful fans produce a cushion of air underneath the flexible "skirt" of a hovercraft. This lifts most of the craft above the water, reducing water resistance (a type of friction) that slows down most boats. Propellers, like those on a plane, push the hovercraft forward.

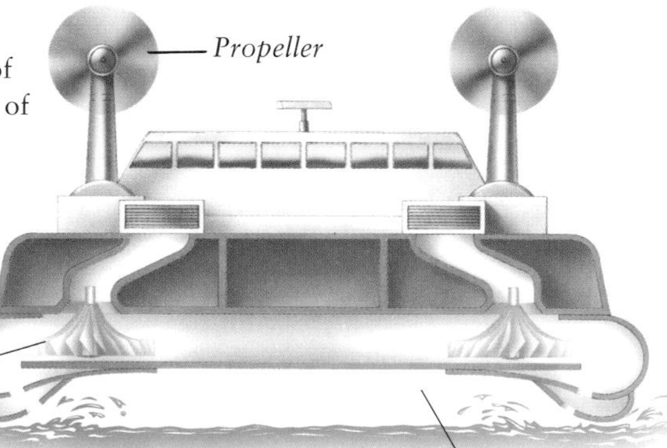

Hovering train

The motors or engines of a train are needed to fight against friction (rubbing) between the wheels and the axles. But some trains use powerful magnets to lift, or levitate, them above the track. These maglev (magnetic levitation) trains need no wheels, and are very quiet.

Up and down

All objects in a liquid experience upthrust—an upward force. If the upthrust is greater than the object's weight, the object floats. If it is less, the object sinks. A submarine crew controls the weight of the craft by varying the volume of air in the sub's buoyancy tanks.

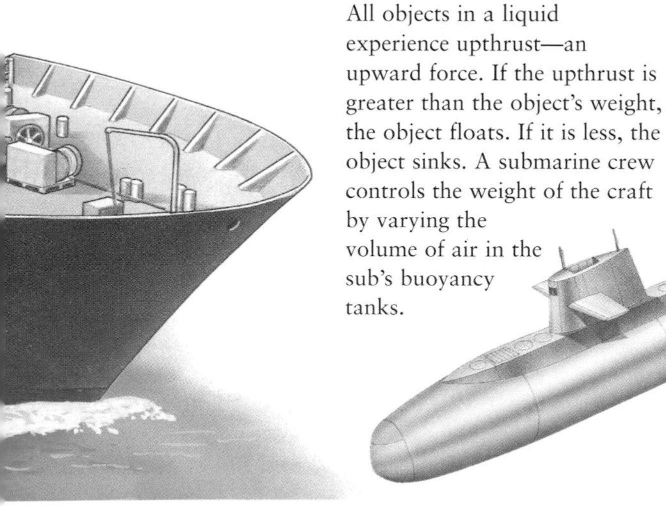

Rudder (steers left or right)

Hydroplane (steers up or down)

> **BOAT AND TRAIN FACTS**
>
> - A double-hulled boat is called a catamaran. A triple-hulled boat is called a trimaran.
> - The speed of a boat through the water is normally measured in units called knots. One knot is one nautical mile per hour. A nautical mile is 1.15 miles (1.85 km).
> - The first railroads carried coal out of mines in horse-drawn carriages more than 200 years ago.

AIRCRAFT

There are four main forces that affect an airplane. The most important is the upward force called lift. This is produced by the wings of the aircraft as they move quickly through the air. It must be greater than the second main force, which opposes lift—the downward pull of gravity that gives the aircraft weight. The third main force is thrust, that pushes the airplane forward through the air. Thrust is produced by propellers or jet engines. It must overcome the fourth force, which opposes thrust. This is drag, also called air resistance. It is a type of friction caused by air rubbing against the plane as it moves along.

Most aircraft wings have flaps that can be raised or lowered to alter the flow of air. These are especially important at lower speeds, when the airplane is taking off or landing. A helicopter uses the same idea to make its lift, but in this case, the wings are its rotor blades, which spin around very quickly. The angles of the blades can be changed, to make the helicopter hover in one place without moving, or travel backward or forward.

Landing wheels fold away after take-off to reduce drag

Propeller (air screw) produces thrust

Cabin for pilot and passenger

Light aircraft

The body of an airplane is called the fuselage. Most small planes—called light aircraft—are pushed through the air by propellers turned at high speed by powerfu engines. There is a variety of moveable control surfaces on the airplane. Ailerons on the main wings help an airplane to roll or bank (tilt to the side) into a turn. The rudder on the airplane's fin (tail) is usec to steer left or right. The elevators or the tailplane make the plane climb or descend.

Lighter than air

Some aircraft, such as hot-air balloons, are called lighter-than-air craft. They are buoyant, in the same way as an object that floats in water. They float because the main part or "envelope" of the balloon weighs less than the same volume of normal air. So the envelope floats upward in normal air, like a bubble in water. In a hot-air balloon, a burner heats the air in the envelope. The heated air expands so that there is less air inside the envelope than before, and the balloon can then rise because it weighs less. Balloon flights are normally made during the morning and evening, when the air is cool. This produces the greatest difference between the air inside and outside the envelope, giving the greatest lift. Airships, on the other hand, are filled with a lighter-than-air gas called helium.

No power

A hang glider is not powered by a propeller or a jet engine. Its wings do not produce lift in the same way as a powered airplane. So a hang glider is more like a steerable parachute than an airplane. It always sinks slowly compared to the air around it. However, if the pilot can steer it into a column of rising air, called a thermal, the hang glider can rise compared to the ground—even though it is still sinking in the air around it.

Envelope

Montgolfier balloon on its first flight

Full of hot air

More than 200 years ago, people first experimented with hot-air balloons. Two French brothers, Joseph-Miche and Jacques-Etienne Montgolfier, made a balloon that rose into the air on November 21, 1783 in France, in fro of an astonished crowd. The balloon with its two pilots, Pilatre de Rozier and the Marquis d'Arlandes, rose about 3,000 feet (900 m) into the air, and sank back again gently ten minutes and 5 miles (8 km) late

Rotor blades

The rotor blades of a helicopter produce lift as they turn. The blades can be tilted—either individually or together—to steer the helicopter, or to rise, hover or descend. A system of control bars at the center of the rotors controls the angle, or pitch, of the blades. Because the whirling blades produce lift as well as controlling direction, helicopters are known as rotary-wing craft.

Air brakes slow speed of plane for landing

Flaps increase area of wing for more lift at low speeds

Curved upper wing surface gives airfoil section

Aileron controls banking or roll of plane, for banked turns

Jet engine on pylon mounting

Rotor blade

Rotor head tilts blades to control helicopter's flight

Drive shaft from engine

Over and under

As a wing moves through the air, it splits the air in two. Air that passes over the wing moves faster than air that passes below it, because the upper surface is more curved. This shape is known as the airfoil section. The fast-moving air produces less pressure on the upper wing surface than the slower-moving air underneath. So there is an overall upward push—lift. The amount of lift depends on the speed the airplane moves, the area of the wings, and the curvature of the airfoil section. Heavy airplanes, such as jumbo jets, need large wings and powerful jet engines to push them through the air quickly enough.

Rudder

Fin

Engine in nose cowling

Aileron

Fuselage

Main wing

Elevator

Tailplane

Main parts of a light aircraft

Light aircraft taxis to take off position

Taking to the skies

There has been a tremendous increase in air traffic over the past 50 years. Most large cities now have airports, and air traffic controllers use sophisticated radar to track the position of airplanes in the sky, in an attempt to prevent accidents. At a busy airport there is one movement—a plane taking off or landing—almost every minute. Usually movements are fewer or stop at night, to reduce noise pollution.

Making Lift

How can you make a piece of paper rise without touching it to push it upward? Cut out a strip of paper about an inch (2.5 cm) wide and 6 inches (15 cm) long. Hold one end between one finger and the thumb of each hand, just in front of and below your bottom lip. Now blow over the top of the paper, and you will see it lift into the air. As you blow, you are creating fast-moving air over the paper, like the air passing over the top of a wing. The still air underneath the paper pushes upward more than the faster-moving air above the paper pushes downward, lifting the paper.

COMMUNICATIONS

THERE ARE MANY different ways of communicating in the modern world. People within speaking distance use words and gestures. But what about telecommunications—getting a message to someone far away? In the past, flags and smoke signals were used to pass information over large distances. Today, we can use the telephone. It is an instant way to communicate, and has a microphone that converts sound into electrical signals. Television and radio are other ways to communicate. Television, radio, and telephone signals pass along metal wires or fiber-optic cables. Printing—in books, magazines, newspapers, and letters—is another way of communicating. Today, the written word can be sent instantly across the world, from computer to computer. The Internet is a vast, worldwide network that allows people on opposite sides of the globe to exchange information between their computers. Information —words, sounds, and pictures—must be digitized (see opposite) before it can be stored, transmitted or received by computers. In the near future, most communication will be digital, allowing even easier exchange of information.

ALEXANDER GRAHAM BELL
(1847-1922)

In March 1876, Scottish inventor Alexander Graham Bell sent the world's first telephone message to his assistant, Thomas Watson: "Mr Watson, come here, I want you!" Bell was calling for help to Watson in the next room at his workshops in Boston, Massachusetts. Other researchers, including Thomas Edison, were trying to develop the telephone, and had some success. However, Bell became famous because he was the first to apply for a patent (a document that claims an invention as yours).

Bouncing back from space

Satellites have revolutionized the way we communicate. They circle the Earth far above the ground, and pass signals from one place to another that may be on the opposite side of the world. Satellites are covered with solar cells, which convert sunlight to electric current. This is used to power all the equipment in the satellite. Satellites are used for many different purposes. Comsats (communications satellites) allow us to make telephone calls to people far away, or see television pictures made in other countries. The signals from the ground are sent up to a satellite and bounced back down to Earth to receiving "dishes". These collect the signals, and pass them to your telephone or television set. Other satellites send information down to Earth, about weather, about other countries, and about space itself.

A satellite sends signals on the downlink to a mid-oceanic island

Comsats

Most communications satellites orbit the Earth more than 21,000 miles (35,000 km) above ground. At this height, they orbit once every 24 hours, so that they appear to stay in a fixed point in the sky. Radio signals from Earth are picked up (received) by dishes on the satellite, made stronger (amplified) and sent (transmitted) to receiving dishes in a different part of the world.

The Internet

Perhaps the most exciting and convenient way to communicate is on the Internet. It is a huge network of interconnected computers. Most people connect their computers to the Internet using modems that convert digital signals into analogue signals (right) that can pass along ordinary telephone cables.

Telephone signals

Inside the mouthpiece, a microphone produces an electrical signal that is a copy of the sound of your voice. The signal passes down cables, through exchanges and to the earpiece of the other telephone, where a loudspeaker produces the sound.

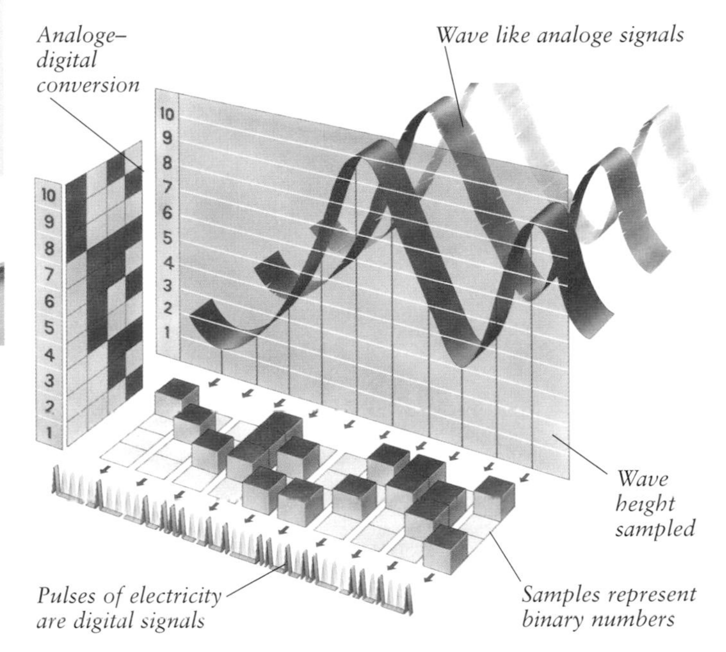

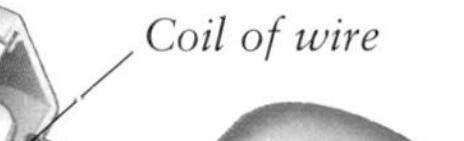

The digital world

Microphones produce continuously varying electric currents that are copies of sounds they pick up. These wave like signals are analoge. They must be digitized—represented as a series of on and off pulses—before they can be used by computers or sent through modern digital telecom networks. This is done by sampling the analoge signal thousands of times each second, and representing the value of each sample as a binary digital number.

On the move

Mobile telephones allow people to stay in touch even when they are not at home. The signals to and from the telephone are carried by radio waves to a local receiving or cellular station, which is connected to the normal telephone network.

Tomorrow's world

In the near future, most telecommunications equipment will send and receive huge amounts of digital information through the fiber-optic cables of ISDN—Integrated Services Digital Network. This will give people access to the Internet, digital television, and radio, and a videophone which will allow them to see the people they talk to.

INDEX

ACKNOWLEDGMENTS

The publishers wish to thank the following artists who have contributed to this book:

Julian Baker, Kuo Kang Chen, Andrew Clark, Chris Forsey, Jeremy Gower, Gary Hincks, Sally Launder, Janos Marffy, Terry Riley, Guy Smith, Sue Stitt, Darrell Warner, Mike White.

The publishers wish to thank the following for supplying photographs for this book:

Page 9 (C) David Parker/Science Photo Library; 15 (T) Alfred Pasieka/Science Photo Library, (C) Novosti Press Agency/Science Photo Library; 26 (B/L) David Parker/Seagate Microelectronics Limited/Science Photo Library; 27 (C) Dixons; 30 (L) Mercedes Benz; 31 (C,T/R) Ford Motor Company; 33 (T,C/R) Brian Morrison; 34 (B) Dover Publications; 35 (B/L) Stansted Airport External Relations Department.

All other photographs from Miles Kelly archives.